Thinking Ecologically

Thinking Ecologically

The Next Generation of Environmental Policy

Edited by Marian R. Chertow

and Daniel C. Esty

Yale University Press   New Haven and London

The Yale Fastback series is designed to provide timely reports on critical issues of the day. Produced on an expedited schedule, Yale Fastbacks are issued simultaneously in cloth and paper in order to reach the broadest possible audience.

Designed by Sonia L. Scanlon

Set in Bulmer type by Ink, Inc., New York, New York

Printed in the United States of America

Library of Congress Cataloging-in-Publication Data
Thinking ecologically: the next generation of environmental policy/edited by Marian R. Chertow and Daniel C. Esty.
p. cm.
Includes bibliographical references and index.
ISBN 0-300-07301-1 (cloth).—ISBN 0-300-07303-8 (pbk).
1. Environmental policy. 2. Environmental protection. I. Chertow, Marian R. II. Esty, Daniel C. HC79.E5T47
1997 363.7—dc21   97-14996 CIP

A catalogue record for this book is available from the British Library.

The paper in this book meets the guidelines for permanence and durability of the Committee on Production Guidelines for Book Longevity of the Council on Library Resources.

10 9 8 7 6 5 4 3 2

For Sarah, Elana, Thomas, Joy, and Jonathan,
who are the next generation

Contents

Acknowledgments

This book emerged from the work of Environmental Reform: The Next Generation Project, sponsored by the Yale Center for Environmental Law and Policy. Beginning with a conference celebrating the twenty-fifth anniversary of Earth Day in 1995, the project evolved from discussions we held with a group of Yale colleagues dedicated to thinking through a vision of the future. Affectionately known as the Rump Group, Reid Lifset, Jane Coppock, Brad Gentry and William Ellis not only provided stimulating discussion but also devoted their time and expertise to every stage of the project.

The basic intellectual framework of the Next Generation Project was enriched by the contributions of a diverse external advisory board unified by their experience and wisdom. Our thanks and appreciation go to Frances Beinecke of the Natural Resources Defense Council, Joan Z. Bernstein of the Federal Trade Commission, John Bryson of the Southern California Edison Company, Leslie Carothers of United Technologies, William Ellis of the Yale School of Forestry and Environmental Studies, Thomas Jorling of International Paper, Fred Krupp of the Environmental Defense Fund, Jeffrey Lewis of the Heinz Family Foundation, Thomas Lovejoy of the Smithsonian Institution, Paul Portney of Resources for the Future, Stephen Ramsey of General Electric, William Reilly of the Texas Pacific Group, Robert Repetto of the World Resources Institute, Edward Strohbehn, Jr., of McCutchen, Doyle, Brown, and Enersen, and Victoria Tschinkel of Landers and Parsons.

We first test-drove our ideas with the graduate students of the Next Generation Seminar at Yale in the fall of 1995. Students from the Yale School of Forestry and Environmental Studies (F&ES), the Yale Law School, and the Yale School of Management brought the passions and politics of this generation to the task of plotting a course for the future. We were gratified by their insights and their hard work on term papers. Our thanks go to class members Saleem H. Ali, Jamie Art, Tom Ballantine, Todd R. Campbell, Joe DeNicola, Andre Dua, Dave Galt, Martha Gray, Gina Gutierrez, Liza Hartmann, Madeline Kass, Chris Lotspeich, Patrick Martin, Ted McCarthy, Rob Naeser, Astrid Palmieri, David Pinney, Andreas Richter, Yasuko Segawa, Shauna Swantz, Cristin Tighe, Antoinette Wannebo, Debra Weiner, Doug Wheat, and Sanghyun Woo, and also to Dini Merz, whose work enriched the course and the project.

During that time we identified "module leaders"—experts in a range of fields—who led daylong symposia on the fourteen topics we selected as key components of next-generation law and policy. We brought in some 250 people for what turned out to be uniformly interesting, lively, and valuable sessions in the winter and spring of 1996. Each symposium became the basis for a chapter in the book. The appendix lists all those who participated, and we owe them our deep thanks for their individual and collective contributions. The module leaders and authors are also listed at the end of the book. We greatly admire their talent and perserverance. Our students worked tremendously hard—Steve Dunn and Mike O'Malley proved to be jacks-of-all-trades. We also thank Kate Bickert, Duenna Chris-Anderson, Lisa Kamemoto, and last, but far from least, Jennifer Thorne.

Over the summer of 1996, the authors of the fourteen chapters worked very closely with us to produce the first draft of the manuscript. Josh Slobin and Matt Gubens of Yale College were especially helpful in reviewing and researching a number of topics. Raj Patel edited a working paper on agriculture and environment emanating from the agriculture module. Marge Camera at the Law School and Anne Wallis at the School of Forestry and Environmental Studies worked on tight deadlines with great patience all though the project.

A number of people provided us with especially helpful thoughts, comments, and suggestions on the direction of the project. We want to thank in particular Marcia Aronoff, Larry Boggs, Marianne Ginsburg, Deborah Gordon, Robert Perry, Jane Polin, David Rejeski, and Lloyd Timberlake.

A two-day symposium in September 1996 was our chance to reconvene the module participants to review the individual chapters, to hear how the overall project was taking shape, and then to announce our findings to the larger community. The second day included very thoughtful talks by Gov. Christine Todd Whitman of New Jersey, Frank Loy, chairman of the League of Conservation Voters, and best-selling author Philip Howard. The open discussion sessions with Next Generation authors greatly strengthened the individual chapters that later emerged. Linda Bergonzi King and Cheryl Volk helped us film and produce a Next Generation video, assisted by Richard Payne, Andrea McQuay, and Carol Leonetti. Our thanks go to Larry Rogero, Michelle Garland, and a host of others who helped organize the symposium. At that time and throughout the process we benefited from the strong support of Yale Provost Alison Richard, Law School Dean Anthony Kronman, and Forestry and Environmental Studies Dean Jared Cohon. Cynthia Atwood and the late Gary Fryer of Yale's Office of Public Affairs helped us with promotion and media efforts.

Preparation of an edited volume is a complex task. Philip Siekman was our principal editor and an ace e-mailer. Thanks as well must go to Tracy Benedict, Jason Brown, Andrew Spejewski, and Jennifer A. McTiernan for helping us push the many pieces of this project forward. We are grateful to Ginger Barber and, at Yale University Press, to Tina Weiner and John Covell for their active guidance.

Publishing a book is not the end for this far-flung project. We are disseminating the Next Generation message and testing its merits in many workshops and policy forums across the United States and abroad. All of our funders have made possible the kind of collaborative process needed for policy thinking. We thank the Association of American Railroads, Avina Foundation, Bechtel Foundation, ERQ Educational Foundation, Geraldine R. Dodge Foundation, GE Fund, General Electric Company, the German Marshall Fund of the United States, Hughes Foundation, and the McKnight Foundation.

On the very day our first grant was announced we had the wisdom to hire Janet Testa to be the project coordinator, and it would be too simple to say that we couldn't have done it without her. Fourteen workshops, two major conferences, and a manuscript were completed in a very short period, thanks to her warmth, commitment, and professionalism. Finally, partnerships start at home, and everything we accomplish reflects the love and strength of our spouses, Elizabeth Esty and Matthew Nemerson.

Thinking Ecologically: An Introduction

Daniel C. Esty and Marian R. Chertow

Thomas Jefferson observed more than two hundred years ago that every generation must reinvent the institutions of society to serve its own needs. A generation has passed and much has changed since Earth Day 1970, which awakened so many Americans to environmental issues and which might be viewed as the starting point for the modern era in environmental law and policy. *Thinking Ecologically* looks back at what this "first generation" accomplished and forward to where the next generation of policies should go.[1]

Like nature itself, the size and shape of environmental problems are constantly evolving. Twenty-five years ago, we faced the challenge of cleaning up rivers so contaminated that one (Cleveland's Cuyahoga) even caught on fire. Air pollution in Los Angeles was so bad that motorists couldn't see three traffic lights in front of them. Toxic waste leached from unlined dumps into schoolyards and basements. Today, we are trying to sort out the long-term effects on our climate of the atmospheric build-up of carbon dioxide and other greenhouse gases. We must consider the potential environmental impacts of genetically modified organisms. And we are trying to understand the risk of exposure to trace residues of pesticides that might disrupt endocrine cycles within a human body. As our targets change, so too must our responses.

In charting a new course for environmental policy, we build on a firm foundation: the American people value and vote for clean water and air, safe disposal of wastes, and the preservation of parks and other special public spaces.[2] In poll after poll, some 80 percent of the respondents indicate that they consider themselves to be environmentalists. Although support for environmental protection is broad, the public's enthusiasm for ever-increasing environmental spending is not unbounded. Indeed, over the last several years, environmental policy has become a bitter battleground, often dominated by extreme points of view. On one hand, deregulators have suggested that we eliminate environmental programs wholesale. On the other hand, some environmentalists have acted as though no improvements in the current structure of American environmental policy were necessary or possible.

Thinking Ecologically seeks to find not only the middle ground but higher ground. It starts with the premise that the flurry of environmental activity that emerged in the late 1960s and resulted in the enactment of a broad range of federal environmental laws in the 1970s moved us forward considerably. The National Environmental Policy Act, the Clean Air Act, the Clean Water Act, the Toxic Substances Control Act, the Resource Conservation and Recovery Act, and more than a dozen other less well known statutes focused America's attention on a range of important public health and natural resources concerns. To a large extent, these laws worked. Our air and water are cleaner. Significant reductions in pollution from big factory smokestacks and effluent pipes have been achieved.

But the prospects for further progress on the same path are limited. Even where existing policy strategies and tools advanced the cause of environmental protection, there is no guarantee that the same line of attack will provide comparable success in the future. Many of today's environmental problems are different from those tackled over the last several decades. Harms such as ozone layer depletion, climate change, or endocrine disruptors are less plainly apparent, more subtle, and more difficult to address than the black skies or orange rivers of the 1960s. Similarly, some of our residual environmental harms represent unresolved problems of the past—automobile exhaust, agricultural runoff, and the loss of habitat to suburbanization—that cannot be solved by clamping down on the emissions of the few thousand largest factories in America.[3] Instead, we must address the effects of thousands of smaller firms and farms whose releases are individually small but cumulatively very large. We must also try to affect the choices of 265 million Americans whose decisions about what to buy, where to live, how much to drive, what to throw away, and where to shop profoundly shape the quality of our environment.

Whether the harms are new or old, the call for creative thinking and fresh approaches should not be viewed with alarm. When asked recently to evaluate the environmental law that he had helped to create as an attorney with the Natural Resources Defense Council in the 1970s, John Bryson, now CEO of

Southern California Edison, observed: "It is not surprising that we got some things wrong. We had no models to follow. What is surprising is that we got anything right."[4] It is with appreciation for what has been accomplished and openness to the opportunity that now exists to refine and update our policies that this book proceeds.

In embarking on a program of environmental reform, it is important to recognize that the American public's overarching goals with regard to environmental protection have remained relatively constant. The vision, articulated in the Clean Water Act more than two decades ago, of lakes, rivers, and streams that are fishable and swimmable remains valid today, as does the Safe Drinking Water Act's call for public water supplies free from harmful contaminants.[5] Thus, while differences in values lie at the heart of a number of environmental policy controversies, a basic commitment to environmental protection as a central element of quality of life is widely shared across the spectrum of U.S. citizens. What is contested is how to move from *first-order* environmental goals—clean air and water and basic protection from carcinogens and toxic exposures—to *second-order* environmental preferences that can be translated into action plans for government, industry, and individual behavior.[6]

This book seeks to reconfigure and reinvigorate the environmental policy debate in America. The ideas and recommendations are the product of a two-year environmental reform initiative undertaken by the Yale Center for Environmental Law and Policy. In the Next Generation Project's two major conferences and fourteen workshops, hundreds of experts from around the world shared their understanding of the successes and failures of our current strategies and their ideas on the direction future policies should take. The contributors to the Next Generation initiative go beyond the usual cast of Washington players. They include dozens of individuals who have real-world experience both in making decisions that affect the quality of the environment and in thinking about how these choices might be made differently and better. The pool of experts includes highway planners, city managers, hospital administrators, farmers, environmentalists, consumer advocates, business people, lawyers, academics, representatives of international organizations, planning and zoning commissioners, and scientists, as well as federal and state environmental and natural resource officials.

From Environmentalism to Ecologicalism

The generational perspective that we have taken is deliberate. The environmental efforts of the past inevitably set the stage for the policies of the future. They represent a legacy—both positive and negative—upon which to build.

The environmentalism of the last twenty-five years was largely pollution-based and law-driven.[7] It often looked disapprovingly at human activities and economic growth because of their harmful pollution side effects, which were thought to be inescapable. Political headway was made by identifying crises, from Rachel Carson's "silent spring" to the toxic stew of Love Canal, and finding villains on whom to pin problems. In this regard, the environmentalism of the 1960s and 1970s was confrontational in style and polarizing in practice. It can be no surprise that the central policy tool was a burgeoning set of "command and control" mandates.

Although the statutory requirements and legal test cases of the 1970s and 1980s brought improvements on a number of fronts, this approach to environmentalism has limitations that are now evident. It compartmentalized problems by environmental media—air, water, waste—and set up a complex structure of separate (and sometimes conflicting) laws and very detailed and often rigid regulations to deal with each new problem uncovered. It encouraged litigation, created incentives for moving with deliberate speed and no faster, and implied a level of absolutism in pursuit of environmental purity that cannot be squared with the public's express and tacit desire for other social goods.

Thinking Ecologically argues for a next generation of policies that are not confrontational but cooperative, less fragmented and more comprehensive, not inflexible but rather capable of being tailored to fit varying circumstances. We see a need for a "systems" approach to policy built on rigorous analysis, an interdisciplinary focus, and an appreciation that context matters. Fundamentally, we seek an *ecologicalism* that recognizes the inherent interdependence of all life systems. This demands, on the one hand, an expanded view of human impacts on the natural environment going beyond pollution to address more subtle, unpredictable, and harder-to-value problems such as habitat destruction, loss of biodiversity, and climate change. On the other hand, it requires an appreciation of the connectedness of all life systems, including human advancement. This focus on linkages and an ecological perspective leads to a more benevolent view of human activities and a belief in the possibility of sustainable development.

The stress on more comprehensive, integrated policymaking is not new. Harold Lasswell argued almost a half century ago for a "policy sciences" approach to public decisionmaking that is contextual, problem-oriented, multidisciplinary, self-aware, and focused on understanding human values.[8] Such a perspective has been advanced more recently by scholars and analysts such as Garry Brewer and Peter deLeon, who updated this view in the 1980s, and Ronald Brunner and Tim Clark, who have applied the more comprehensive approach of the policy sciences to natural resource problems in the 1990s.[9]

To some observers, the call for more comprehensive perspective and greater attention to interconnectedness harks back to innumerable pleas for such virtues in the 1960s. But integrated and broad-scale thinking is possible today in ways that were unimaginable a generation ago. To begin with, we have a generation's worth of policy practice and experience to build upon. More important, advances in information technologies make the amassing, assessing, and simultaneous processing of vast quantities of data not just conceivable but ever easier. Thus, although entirely comprehensive policy analysis will remain elusive and environmental policymaking contains an irreducible dimension of political value judgment, our capacity to be more all-encompassing and integrative has evolved dramatically. We are positioned, as Lasswell urged, to understand the relationship between the parts and the whole of a system and to blend knowledge *of* the policy process (theory) with the knowledge *in* that process (practice).

An emphasis on interconnectedness has several implications. First, policy thinking needs to be infused with knowledge from outside the environmental sphere about real people's real lives. Failing to understand the competing desires that citizens everywhere have for a cleaner environment *and* other things—mobility, economic growth, jobs, competitive industries, and material comforts—leads to policies that are out of sync with the people whose lives they are meant to serve and diminishes the prospects for winning the public and political support necessary to effect change.

Second, environmental protection must become everybody's business. At one level, this requires more activism on the part of citizens in their roles as local leaders and as individual consumers—factoring environmental considerations into the choices they make for their families and their communities. But at another, perhaps more important, level, all Americans can be drawn into the environmental policy process by ensuring that the prices they pay for goods and services reflect the full costs of any public health or ecological harms associated with these purchases. In this way, market forces ensure that every citizen becomes a constructive environmental decisionmaker, even if he or she never realizes it.

In addition, our vision of environmental decisionmaking needs to extend beyond the choices made by EPA or state department of environmental protection officials. The decisions of mayors, transportation system designers, city planners, farmers, wholesale distributors, and business people of all stripes profoundly affect the quality of our environment. They must come to recognize and embrace their responsibilities and opportunities as environmental decisionmakers.

Finally, an emphasis on connectedness and a more comprehensive policy perspective requires that we be rigorous in ensuring that our policies deliver "environmental value"—effective and efficient protection across the spectrum of public health and ecological risks we face. First-generation policies often emphasized technology "fixes" and downplayed concerns about economic efficiency. Where problems loomed large, this approach promised (and delivered) relatively quick results. But we now face diminishing returns to this strategy and a need to be more deft in our policy interventions. Long-term public support for environmental investments depends on the programs delivering good value on the money (public and private) put into them. Simply put, we must get the maximum bang for society's inevitably limited environmental buck.

Institutional Realignment

One of the central challenges for environmental policymakers is to keep pace with the important elements of institutional realignment that are occurring in society. Notably, the role of government is narrowing, the private sector's responsibilities are broadening, and nongovernmental organizations, from think tanks to activist groups, are increasingly important policy actors.

The corporate world is not monolithic with regard to environmental performance. Some companies take environmental stewardship very seriously and are among the most progressive forces for environmental progress in the world. Other companies continue to pollute with abandon and to seize public resources (water, air, land) as though they were free for the taking. If the next generation of environmental policies is to be successful, separating the leaders from the laggards in the business world will be essential. With limited resources available, governments must target their enforcement activities on those whose performance is not up to par.

Redefining the role of government is, perhaps, the central question of our age. Both major U.S. political parties have called for a smaller role for the federal government. The environmental dimension of this challenge must be tackled head on. Leaner and more efficient environmental programs are a necessity. But the need for government intervention will not disappear. Quite to the contrary, the triumph of market economics makes all the more clear the need for laws and rules to ensure that pollution harms do not go unaccounted for. In fact, pollution represents what economists call an *externality*—a cost that can be pushed out a smokestack or effluent pipe, or otherwise unfairly dumped onto others. Unless government acts to *internalize* such harms— requiring polluters to control their emissions or pay for the harm they cause—market failure and diminished welfare will result.[10]

Government intervention is similarly necessary to avoid the tragedy of the commons, where individuals acting in their own rational self-interest consume common resources, such as air or water, at rates that are unsustainable if replicated by all.[11] In some cases, clarifying who has a right to do what—that is, spelling out who has the "property rights"—will lead to workable solutions. But in many cases, the government must establish and enforce the bounds of acceptable environmental behavior to ensure optimal social welfare and the efficient functioning of our market economy.

Recognizing that there is an important role for government in protecting the environment does not answer the question of what level of government should act. In fact, "devolution," as much as "deregulation," has been the rallying cry for some recent environmental reformers. Their concerns about overly rigid policies dictated from Washington that do not match local needs and circumstances often strike a responsive chord. Toxic waste cleanups should be done in ways that meet community interests. And rivers are better managed at the watershed level than by officials hundreds or thousands of miles away. But decentralization is not always the right policy answer. A more refined line of analysis suggests that environmental policies should match the scale of the problems to which they are addressed.[12] Thus, although some of our programs would best be moved to local communities, in other cases (such as ozone layer depletion), the scale of the harm requires intervention at a more overarching level.[13]

Environmental policy must also be updated to reflect the evolution of "civil society." Americans have long been participants in a wide range of nongovernmental organizations (NGOs), clubs, teams, and religious groups. NGOs play an especially important role in the environmental domain. Individuals united by a common commitment to an environmental resource, whether the Chesapeake Bay, California's Central Valley, the Gulf of Mexico, or even the Amazonian rain forests, can join together, wherever they may live, to engage the political process across traditional geographic boundaries.

A Few Caveats

Some of the proposals advanced in this book represent new approaches to environmental policy. Many of the concepts, however, are not novel but rather emerged from our search across the country and around the world for innovative approaches to environmental protection that had not been widely or consistently translated into the policy domain. *Thinking Ecologically* is not, therefore, a precise game plan for policy reform but rather a menu of ideas that we hope will stimulate creative thinking at all levels of society about how to achieve environmental progress. We offer no simple answers, and indeed there are none. The environmental problem is not a single issue but rather a diverse and complex set of public health, habitat conservation, and resource management concerns—requiring an equally diverse set of responses, not only from government at all levels but also from businesses, nongovernmental organizations, and individuals. The next generation of environmental policymakers will require a bigger toolbox than the last generation, and responsibility for implementing the resulting policies must be placed in broader hands.

We recognize, furthermore, that we have not addressed all of the issues that might be considered important to successful environmental policy reform. We have no chapter on environmental equity, nor one on environmental education. The important questions of who benefits from environmental policies and who suffers from insufficient (or excess) attention to environmental harms are, however, addressed in various places throughout the volume. Similarly, we think that this entire project is about environmental education. Taken as a whole, we believe *Thinking Ecologically* will provide an important step forward in public understanding about the issues and the trade-offs inherent in environmental policymaking.

Although there is no bright line separating international from domestic environmental protection, we have focussed primarily on U.S. concerns. Nevertheless, several chapters address global issues, and we believe that much of what we discuss in the U.S. context will have resonance abroad.

Foundations for the Next Generation of Environmental Policy

This volume begins with a series of chapters that introduce a set of new foundations for ecological thinking and policymaking. In chapter 1, Professors Charles Powers and Marian Chertow examine the current fragmented structure of U.S. environmental policy, which often leads to tunnel vision,[14] the cycling of pollution, and regulatory processes that are too narrowly targeted. We "fix" our air pollution problems with scrubbers that create a sludge that becomes a land disposal issue which, if improperly handled, may run off into streams, becoming water pollution. Powers and Chertow introduce, as the centerpiece of a next-generation systems-oriented environmental policy, the concept of "industrial ecology," which allows us to go beyond "pollution prevention" toward more comprehensive, life-cycle approaches to environmental protection. They also offer a transition strategy for moving toward a systems approach to environmental protection.

Some of the bitterest fights over environmental policy in the last few years have arisen in circumstances where survival of a single species or an environmental resource such as a wetland has been pitted against jobs and economic growth. In chapter 2, Professors John Gordon and Jane Coppock advocate an ecosystem approach to environmental policymaking that seeks to optimize both ecological protection and economic growth. They argue that ecosystem management offers a more integrative—ecological—analytic framework that can help decisionmakers channel development and manage across the multiple values and resources found on any piece of land. Ecosystem management, they observe, is not only an analytic tool but also represents a cooperative process through which divergent interests in the use of land and variations in the values that individuals and communities place on resources can be accommodated.

In chapter 3, with a sweeping historical perspective, Prof. Carol Rose reclaims "property rights" as a concept that supports sound environmental policies. Recent debates over the role of property rights have often become tangled in anti-environmental rhetoric. Rose notes that the five-hundred-year Anglo-American legal tradition lodges property rights not only with landowners but also with their neighbors and the community at large, who are entitled not to have harms spilled onto them. But she adds that it may be useful, as society's definition of "harm" evolves and we refine our policy focus, to find some accommodation for landowners whose plans and traditional activities are disrupted by changed conditions or new scientific knowledge. Facilitating the transition to a new set of environmental expectations and requirements helps to avoid the political backlash that may otherwise ensue.

In chapter 4, John Turner and Jason Rylander of the Conservation Fund take on the disconnect between environmental policy, which is largely set at the state or federal level, and land use decisionmaking, which largely occurs at the local level. For most of the last twenty-five years, environmental policymakers have acted as though these two realms of decisionmaking were entirely unconnected. But, obviously, the cumulative effect of local land use decisions is the prime determinant of environmental quality more broadly. We must, they argue, identify new ways to link up local, state, and federal decisionmaking. And we need to draw the private sector, nongovernmental organizations, and individual citizens into our quest for better environmental protection and resource management.

Our present structure of environmental law and policy focuses almost entirely on the activities of manufacturing companies. Yet today, 76 percent of our gross domestic product is in the services sector. Dr. Bruce Guile and Prof. Jared Cohon argue in chapter 5 that we must update our environmental programs and reorient them to this new economic reality. They see the policy transition from a smokestack economy to a services one entailing both challenges and opportunities. Where once people went to downtown shops and more recently to regional malls, Americans are increasingly buying through catalogs. Does overnight shipping from L. L. Bean via Federal Express translate into better or worse environmental outcomes than the past pattern of purchasing? We have not yet even begun to analyze, never mind answer, such questions. The structure of the services economy also creates new points of environmental leverage. For example, when retailing giant Home Depot determines what kind of wood preservatives will be permitted in the lumber it sells, a de facto standard will have been set for an entire industry. The right choice in this case could have profound environmental implications. We see an expanding role for this type of nongovernmental policymaking.

Not only has the domestic economy been radically transformed in the last twenty-five years, but so too has America's role in the world. As Elizabeth Dowdeswell, executive director of the United Nations Environment Programme, and international environmental policy expert Steve Charnovitz point out in chapter 6, our society is increasingly global in scope. Expanded trade implies economic interdependence, and a set of inherently global environmental problems leaves us in a position of ecological interdependence. Dowdeswell and Charnovitz suggest that managing this interdependence will require improved international environmental institutions and better approaches to pollution problems that arise on a global scale.

New and Renewed Tools and Strategies

The shift away from the polarities of the past several decades toward a policymaking model that attends to interconnectedness demands new tools and strategies. For several decades, economists have urged the adoption of market mechanisms rather than traditional command-and-control regulatory strategies, noting the benefits of aligning the economic incentives of polluters and consumers with environmental goals. Yet, as Prof. Robert Stavins and McKinsey & Company consultant Bradley Whitehead spell out in chapter 7, our current regulatory structure still relies largely on command-and-control mandates. Stavins and Whitehead review obstacles and opportunities for a range of market mechanisms, from "green fees" to pollution allowance trading systems to pay-as-you-throw garbage regimes designed to encourage safe and cost-effective waste management. They see a need to celebrate the successes of market mechanisms such as the 1990 Clean Air Act's acid rain trading program, which has helped to reduce the cost of eliminating sulfur from the air by hundreds of millions of dollars, and to expand public appreciation of the benefits of incentive-based policies.

Business has replaced government as the central engine of environmental investment, particularly in the developing world. As business-environment experts Stephan Schmidheiny and Bradford Gentry make clear in chapter 8, next-generation policymakers must find ways to channel private capital flows to ensure investment in environmental infrastructure and adherence to appropriate environmental standards in all new factories and other development projects. The scope of the opportunity cannot be overstated. In 1996, for example, China received about 2.5 billion dollars in *official* (governmental) assistance from all sources including the World Bank, other multilateral development banks, and bilateral aid donors such as Japan and the United States. During that same year, China absorbed more than 40 billion dollars' worth of foreign investment. How this private capital is deployed will have a far greater impact on China's environmental future than the government monies.

Technology innovation offers one of the most promising routes to better and more cost-effective environmental programs. Yet, our current structure of environmental law and policy often deters innovation and retards technology development. Entrepreneur John Preston argues in chapter 9 that part of the problem is a regulatory structure based on technology mandates rather than performance standards. In addition, he identifies a systematic gap in the funding necessary to move new ideas from the early development stage to commercialization. Preston calls for a number of creative strategies for bridging the technology gap and ensuring that U.S. policy promotes innovation and creative thinking.

To the extent that our environmental policies must become more refined and analytically rigorous, we will need better science, data, risk analysis, and cost-benefit studies on which to base policy decisions. Public health authority James Hammitt spells out in chapter 10 both the need for improved analytic foundations for environmental policy and the benefits of a more transparent policymaking process that clarifies where the domain of science ends and that of political judgment begins. He identifies the spectrum of tools that are available and observes that although the limitations of science are real, there is no other foundation for good environmental decisions than credible data and analysis. He also notes that today's policymakers must track a wider range of ecological and public health harms than was the case a generation ago.

Our structure of law and implementing policies must become more ecological if continued progress is to be made on the environmental front, argues E. Donald Elliott, former general counsel of the Environmental Protection Agency. In chapter 11 he calls for a new strategy for environmental regulation that emphasizes "command and covenant." Under this approach, businesses would be required to meet government-determined performance goals but would be freer to determine for themselves how to achieve the established standards. Their success in meeting the requirements would be monitored by independent auditors, similar to those that review the financial statements of companies under Securities and Exchange Commission rules. Limited government enforcement resources could then be targeted toward those failing to live up to their environmental obligations.

Broadening the Reach of Next-Generation Environmental Policymaking

Most of the decisions that effect environmental quality are not made by officials of the EPA or of state departments of environmental protection. A policy model based on interconnectedness allows us to see that environmental results derive in large measure from decisions made in other realms such as energy, agriculture, and transportation. Thus, one of the central thrusts of the next generation of environmental policy must be to redefine the sphere of environmental decisionmakers to include the actors in these sectors. With this vision in mind, the final chapters of *Thinking Ecologically* look at how environmental policy might be made in three critical areas. In chapter 12, former Connecticut Transportation Commissioner Emil Frankel explores the range of environmental tensions created by an American society defined by the automobile. He advocates a variety of innovative approaches, from pay-as-you-drive fees to more "intelligent" highways as a way of reconciling Americans' love of cars and their contradictory interests in both mobility and clean air.

Prof. C. Ford Runge makes clear in chapter 13 that there has been no "first generation" of environmental policy in the agricultural domain. In the farming realm, the environmental steps that have been taken are minor subtexts in a plot that revolves around crop subsidies. He urges that basic economic incentives be brought to bear on farm policy and that the transition away from subsidy-based production be eased by means of a negative pollution tax that would reward farmers who attend to environmental problems.

In chapter 14, Prof. Todd Strauss and Enron Corporation executive John Urquhart review the complex relationship between energy and environmental policies. They observe that energy prices must be made to reflect environmental harms. They see the deregulation of energy markets creating significant environmental challenges as well as presenting some opportunities.

The last chapter of the book offers a vision of what the world might look like in 2020 if the next-generation policies we have introduced take hold. Chapter 15 provides, in addition, a glimpse of how the ideas from the rest of the book could be translated into the fabric of everyday society.

Paul Portney of Washington-based Resources for the Future has argued that the surgery required for successful environmental policy reform must be undertaken with a laser beam, not a chain saw.[15] What is required is the not-so-glamorous job of making dozens of policy refinements. While carrying forward a policy *transition* is not as flashy as executing a policy *revolution*, it is likely to produce more durable improvements to our programs for environmental protection. Although it runs counter to the instincts of most politicians and many in the media, there are no "sound bite" solutions to the challenges we face in the environmental domain.

Given the uncertainties inherent in environmental decisionmaking, regulatory reform must be viewed as a process, not an endpoint. Drawing on concepts such as the focus on continuous improvement from total quality management, environmental policymakers must view their job as one of constant reassessment and refinement as new information becomes available and data on past policy efforts provides a basis for sharpened future responses. At the core of this process is the need to view environmental challenges comprehensively and to build bridges from science to politics, from academic theory to practical policy, and across the gulfs that divide the public and government officials.

Inspiring the American people to support careful, thoughtful, systematic, and enduring environmental reform in a context where the enemy is hard to see and progress is measured incrementally poses a significant challenge. Appropriate and enduring environmental policies rarely emerge from black-and-white visions of reality. Yet finding one's way through the grey area can be slow and excruciating. In establishing an appreciation of the interconnections that must be understood for good policymaking—of the ecologicalism required—we hope this book will have made a contribution to the next generation of environmental policy.

Notes

1. Although we define the period 1970–95 as the first generation of modern environmental law, because it represents an unprecedented burst of national regulatory activity, there were earlier efforts aimed at various conservation, resource, and pollution issues. For a history of America's environmental movement, see Philip Shabecoff, *A Fierce Green Fire* (New York: Hill and Wang, 1993).

2. See Everett Carll Ladd and Karlyn H. Bowman, *Attitudes Toward the Environment* (Washington, D.C.: AEI Press, 1995).

3. In technical terms, we have done a great deal to address the biggest "point" sources of pollution; we have done much less to control "nonpoint" source emissions. These diffuse harms persist in part because they are hard to see, not easily measured or matched to the ills they inflict, and difficult to prevent or control. Of course, the slow pace of progress stems not simply from inadequate policy tools but also, in some cases, from a lack of political will.

4. John Bryson, speech at the Yale Law School, October 1994.

5. We take "fishable, swimmable," and "free from contamination" at face value as broad objectives. Achieving these at any cost, however, is not the spirit of the goals that we wish to carry forward.

6. A wide range of projects are under way that aim at regulatory reform, reinventing the Environmental Protection Agency (EPA), and bringing about other environmental change. See, for example: Aspen Institute, *The Alternative Path: A Cleaner, Cheaper Way to Protect and Enhance the Environment* (Washington, D.C.: Aspen Institute, 1996); Center for Strategic and International Studies, National Academy of Public Administration (NAPA), and the Keystone Center's Enterprise for the Environment Project (report forthcoming, 1997); Competitive Enterprise Institute: Ronald Bailey, ed., *The True State of the Planet* (New York: Free Press, 1995); Environmental and Energy Studies Institute, Leadership Initiative for New Environmental Strategies, "New Strategies—at a Glance" (Washington, D.C., 1996); National Environmental Policy Institute (NEPI), Reinventing EPA and Environmental Policy Working Group, the Unified Statute Sector, *Integrating Environmental Policy: A Blueprint for Twenty-first-Century Environmentalism* (Washington, D.C.: NEPI, 1996), and NEPI, *Reinventing the Vehicle for Environmental Management: Reinventing EPA/Environmental Policy: First Phase Report* (Washington, D.C.: NEPI, 1995); Presidential/Congressional Commission on Risk Assessment and Risk Management, *Framework for Environmental Health Risk Management,* Final Report, vol. 1 (Washington, D.C., 1997); Bruce Yandle, *A Positive Agenda for Environmental Policy* (Washington, D.C.: Progress and Freedom Foundation, April 1996); Debra S. Knopman, *Second Generation: A New Strategy for Environmental Protection* (Washington, D.C.: Progressive Foundation's Center for Innovation and the Environment, April 1996); and Office of the Vice President, National Performance Review, *Accompanying Report of the National Performance Review,* "Improving Regulatory Systems" (Washington, D.C., September 1993) and "Reinventing Environmental Management" (Washington, D.C.: Government Printing Office, 1994).

7. The model employed was that of the civil rights movement, in which sweeping legal requirements were identified and then given detailed content through test cases in a variety of contexts.

8. This approach was described in a book edited by Daniel Lerner and Harold Lasswell called *The Policy Sciences: Recent Developments in Scope and Method* (Stanford, Calif.: Stanford University Press, 1951). Lasswell's *A Pre-View of Policy Sciences* (New York: American Elsevier, 1971) represents the elaborated statement of this vision. In addition, Lasswell worked closely with international legal scholar Myres McDougal of the Yale Law School, and chap. 2 of Lasswell and McDougal's book *Jurisprudence for a Free Society: Studies in Law, Science, and Policy* (New Haven: New Haven Press; Dordrecht, Netherlands, and Boston: M. Nighoff, 1992) adds important perspectives.

9. See Garry D. Brewer and Peter deLeon, *The Foundations of Policy Analysis* (Homewood, Ill.: Dorsey Press, 1983). See also Ronald D. Brunner and Tim W. Clark, "A Practice-based Approach to Ecosystem Management," *Conservation Biology* 10, no. 5 (October 1996): 1–12. The recent call for a "comprehensive approach" to climate change echoes this perspective—see Richard Stewart and Jonathan Wiener, "A Comprehensive Approach to Global Climate Policy: Issues of Design and Practicality," *Arizona Journal of International and Comparative Law* 83, no. 9 (1992).

10. William Baumol and Wallace Oates, *The Theory of Environmental Policy,* 2d ed. (Cambridge: Cambridge University Press, 1988).

11. Elinor Ostrom, in *Governing the Commons: The Evolution of Institutions for Collective Action* (Cambridge and New York: Cambridge University Press, 1990) argues that those who use common pool resources may devise mutually workable rules and institutions for managing the resources even without government intervention. Similar arguments are reviewed in Daniel Bromley, ed., *Making the Commons Work: Theory, Practice, and Policy* (San Francisco: ICS Press, 1992).

12. See Daniel C. Esty, "Revitalizing Environmental Federalism," *Michigan Law Review* 95 (1996): 570–653.

13. Not only are some harms inherently global in scale, but the geographic reach of some issues is bigger than we had previously understood, and the scope of the policy response must be concomitantly larger. For example, recent studies have shown that heavy metals such as mercury can travel thousands of miles through the atmosphere (see William Fitzgerald, "Mercury as a Global Pollutant," *The World and I* [October 1993]: 192), and the latest ozone (smog) transport modeling shows much greater interstate spillovers than current policies anticipate.

14. See Stephen Breyer, *Breaking the Vicious Circle: Toward Effective Risk Regulation* (Cambridge: Harvard University Press, 1993).

15. P. R. Portney, "Chain-Saw Surgery: The Killer Clauses Inside the 'Contract,'" *Washington Post,* 15 Jan. 1995: 23.

part one

Foundations for the Next Generation

Industrial Ecology

Overcoming Policy Fragmentation

Charles W. Powers and Marian R. Chertow

A generation ago the task of environmental protection seemed simpler. Pollution of air, water, and land was the unwanted by-product of economic activity and had to be stopped, typically by very specific directives embodied in strong central government regulation. In contrast, a generation later, we comprehend the underlying issues of environmental protection quite differently. We find that we have only just begun to recognize the interconnectedness of ecological phenomena and to see that our past policy and law have often missed the mark or been unresponsive to new scientific and technical knowledge.[1]

Through the lens of the emerging field of industrial ecology, "economic systems are viewed not in isolation from their surrounding systems but in concert with them."[2] What pollution is and how we respond to it are now seen in context. Waste, for example, does not have to be viewed as a problem if it can be used efficiently by another company as a feedstock. In the broadest sense, environment is seen not as a place removed from the world of human activity but as intrinsic to "industrial" decisionmaking—whether industry is interpreted narrowly as a particular organization or more expansively as the scope of human activity.

Industrial ecology is a systems approach to the environment. It suggests a more comprehensive view of environmental protection than the laws and policies of the 1970s and 1980s, which divided pollution into many separate problems based on categories of places, products, and poisons. This chapter explores how industrial ecology, which has its anchor in science, can become a guide to policy. It examines first-generation environmental regulation, discusses the systems problems encountered, describes why a policy framework based on industrial ecology could overcome some of the dilemmas identified, and offers a transition strategy, also based on systems thinking, that can move us from where we are now to a more coherent environmental policy approach.

Looking Back

In 1970, the seeds were sown for both a comprehensive approach to environmental policymaking and a much more targeted one. The National Environmental Policy Act (NEPA), passed into law that year, took only five pages to make it the policy of the United States government "to create and maintain conditions under which man and nature can exist in productive harmony" and to authorize partnerships and problem-solving collaboration between the public and private sectors to implement that policy.[3] That same year saw passage of the Clean Air Act, which, in the course of hundreds of pages of tightly wound definitions, standards, penalties, and liabilities, focused attention on the problems created for health and welfare in a single medium—air.

In the following years, the broad scope of NEPA was largely forgotten in Congress's rush to introduce specific environmental legislation as the answer to a litany of public concerns. The newly formed U.S. Environmental Protection Agency (EPA) was eager to establish a track record of responsiveness to Congress and the American people.[4] The 1970 Clean Air Act was soon followed (and rivaled in complexity) by the 1972 Clean Water Act. With these precedents, calls for other specific legislative responses arose and a welter of new laws, regulations, liability rules, and "guidance" descended. Little attention was paid to overlaps, gaps, and conflicts between different pieces of legislation.

There were good reasons, initially, to parse the work of environmental protection into air, water, and waste and subdivisions of each category and into separate classes such as pesticides or hazardous materials. Indeed, as we indicate in this chapter, dividing up problems makes them more tractable and accessible. This approach provided a useful starting place for our modern environmental protection efforts, which in fact have been quite successful. By the mid-1980s, the air was purer. Dying rivers again supported life. Toxic contaminants were stopped before they reached aquifers. Pollution was extracted from the places on which the spotlight of environmental regulation was shined.

Still, the piecemeal regulatory approach was problematic. Differentiating pollution by media and class did not per se cause the system to function poorly. Rather, problems arose because such differentiation led to fragmentation. In the words of policy scientist Harold Lasswell: "Fragmentation is a more complex matter than differentiation. It implies that those who contribute to the knowledge process lose their vision of the whole and concern themselves almost exclusively with their specialty. They evolve ever more complex skills for coping with their immediate problems. They give little attention to the social consequences or the policy implications of what they do."[5]

Within the U.S. environmental protection program, there are several categories of fragmentation: by type of pollutant, by life-cycle stage, and by organizational characteristics.

Fragmentation by type of pollutant pertains to how we regulate different contaminants. We know now, as we suspected then, that pollution does not respect legislated boundaries such as air, water, and waste. Sulfur dioxide released into the air, even by a tall smokestack, does not disappear but can come back as acid rain that threatens water supplies. If we trap emissions before they leave the smokestack we create a sludge that becomes a hazardous waste disposal challenge. Fragmented law fails to account for instances where pollution is merely shifted from one place to another rather than reduced or eliminated.

First-generation environmental laws also led to fragmentation among the stages of what we now call the product life-cycle—the chain extending from extraction of materials to manufacturing to distribution to use of products and to their ultimate reuse or disposal. We discovered that regulation centered on a factory's emissions does little to reduce environmental impacts caused when the parts or materials used in that factory have already been produced by suppliers elsewhere. Nor does a facility-centered approach address the environmental problems that a product can cause a firm's customers, whether they are distributors, retailers, or final consumers. Unfortunately, it is often the regulatory structure itself (in this case particularly the Resource Conservation and Recovery Act [RCRA]) which, because it focuses on isolating waste, often precludes the simple and organic exchange and reuse of wastes. These laws severely limit the possibility of recycling or recovering many hazardous materials even if their reuse generates clear environmental benefit.

Organizational fragmentation has proven to be troublesome in two ways: through the means developed to implement regulations and through the problems of organizational culture that followed. Each law for each medium developed different definitions, standards, and approaches to penalties and liabilities as well as different kinds of prescriptiveness, ways to "trigger" regulatory action, and mandates concerning which governmental agency should do what to whom and when.[6] Subsequently, each law developed its own compliance culture: measures, permits, emissions standards, definitions of what is and is not included, and patterns of liability that served to focus people's attention on ways to satisfy specific regulatory requirements rather than on ways to enhance the environment. Calls to "create and maintain conditions under which man and nature can exist in productive harmony" became the stuff of conservationists' speeches, while the nation plunged into the tasks of enforcing, complying with, and litigating over the new sets of laws and regulations.

By the mid-1980s, the system began to show signs of strain. Experts reported significant volumes of hazardous materials in places just beyond the regulatory spotlight—sometimes in pathways to receptors that were just as worrisome as the ones from which the materials had been diverted. A decision to make an investment in a longer- or shorter-term "fix" to achieve compliance was increasingly tied not to potential benefits to the environment but to careful calculation about whether and when a regulatory proposal would pass and whether regulations would change in response to new data. Adding to the reluctance of the regulated community was the awareness that as more and more pollution was cleaned up, the cost of managing or preventing the remaining increments skyrocketed.

Confidence that the web of laws covered the right issues and did so effectively began to wane. Environmentalists pointed out that fundamental, long-term threats to the public's health and the environment—such as habitat loss and pesticide exposures—were being overlooked because policymakers focused on too limited a range of issues and were blinded by fragmented policy to cumulative impacts.

Slowly at first, then at a more rapid clip, situations came to light in which the existing laws appeared to lead to or encourage practices worse than what was being fixed.[7] The perception that some of our laws might even be counterproductive began to grow. Typically, the evidence was not dispositive. In most cases, the scientific community remains split, even today. Do technologies aimed at reducing the volume of ten micron-sized particulates lead to the emission of even smaller and substantially more dangerous ones? Do gasoline additives intended to increase oxygenates to reduce carbon monoxide emissions actually cause harm? Do asbestos rules actually tend to increase fiber levels in schools by encouraging inappropriate removal? Are new industrial facilities located in pristine "greenfields" where they require new, costly, and environmentally burdensome infrastructure to avoid possible Superfund liability expenses that might be incurred by rehabilitating existing industrial locations?[8] Even if the answers to these questions were not always "yes," they were all too often "quite possibly."

Reclaiming a Long-Term Vision

Evolving knowledge of human and ecological systems has helped us understand the multiplicity of interrelationships, their complex interactions, and the long-term nature of the risks they pose. A fragmented approach to environmental policy diverts attention from the careful analysis, integrated perspective, and creative problem solving required to understand and protect public health and natural resources. As the limitations to the first-generation approach have become more and more apparent, environmentalists and industrialists alike have begun to look at the existing laws as impediments to imagination, to common sense,[9] to incentives for broad-based thinking, to efforts to link diverse environmental issues, and to systematic pursuit of environmental and other social purposes simultaneously. The reemergence of more comprehensive thinking revives the overarching goal of NEPA—to create and maintain conditions under which people and nature can exist in productive harmony—which has begun to inspire thinking about a next generation of "ecological" policymaking.[10]

A central thrust of next-generation environmental policy must be to move beyond the regulatory and organizational barriers that single-media, single-species, single-substance and single–life-cycle-stage approaches create to a more holistic and longer-term consideration of environmental threats. It must resist fragmentation, help overcome cultural barriers, and deal with questions of complex interactions between the economic and natural worlds. It must also resourcefully avoid the problems created by categories that are too limited, tools that are too blunt, and thinking that is too narrow. In one way, the ecological perspective we seek represents a return to the beginning of the modern environmental era. In fact, twenty-five years of learning and experience allows us, now, to reopen the door to a more inclusive policy approach.

Industrial Ecology and Policy Progress

In the quest for better policies, the emerging field of *industrial ecology* can provide a beacon for next-generation thinking. Industrial ecology emphasizes a systems view. The use of *ecology* focuses attention on how the natural world works. It highlights the opportunity to look to the natural world for models of efficient use of resources, energy, and wastes. Industrial ecology examines local, regional, and global materials and energy flows in products, processes, industrial sectors, and economies. It focuses on *industry* in two senses. In its broadest meaning, industry refers to all types of human activity. Industrial ecology studies the connection of people and nature, placing human activity in the larger context of the biophysical environment from which we obtain resources and into which we put our wastes. It also sees industries—that is, corporate entities of all kinds—as key players in environmental protection.[11]

Industrial ecology is a new field: the first major colloquium on the topic was held at the National Academy of Sciences in 1991.[12] It has come to embrace several systems concepts and builds on many antecedents.[13] It blends optimism about technological development with keen interest in models from the life sciences and grounding in the systems sciences. It routinely relies on certain tools such as design-for-environment (DFE) and life-cycle assessment,[14] although it did not create them. Still, industrial ecology has proven to be "an effective framework for applying many existing methods and tools, as well as for developing new ones."[15] Although no single declaration "explains" industrial ecology, use of the emerging cluster of ideas grouped under its banner by members of the scientific and engineering communities has proven fruitful at three levels:

Within the firm. Tools such as full-cost accounting and design for environment have proven to be useful ways of drawing together financial and environmental considerations into one system. As AT&T's Brad Allenby observes: "In the short-term, Design for Environment is the means by which the still vague precepts of industrial ecology can in fact begin to be implemented in the real world today. DFE requires that environmental objectives and constraints be driven into process and product design, and materials and technology choices."[16]

Between firms. Crossing firm boundaries has led to sharing of resources such as water, power, and waste across companies in eco-industrial parks.[17] This thread is elaborated by General Motors executive turned Harvard professor Robert Frosch: "The idea of an industrial ecology is based upon a straightforward analogy with natural ecological systems. In nature an ecological system operates through a web of connections in which organisms live and consume each other and each others' wastes. . . . In the industrial context we may think of this as being use of products and waste products."[18] In addition, firms have recognized that their products cross many company boundaries during their life-cycles from design and manufacture to distribution to use to final disposition. When a company such as Duracell or S. C. Johnson decides to "green the supply chain"—that is, demand of their many suppliers that each meet environmental goals—they are using a life-cycle framework for environmental improvement.

Regionally and globally. Tracking flows of material and energy across regions, economies, and the globe illuminates what happens to the constituents of industrial and commercial products.[19] A study by Valerie Thomas and Thomas Spiro of lead in the world economy, for example, plotted the source and use of 5.8 million metric tons of lead consumed annually.[20] It showed how much lead was produced, how much was recycled, and how much was "lost" into the environment. It differentiated "one-time" uses such as lead shot discharged directly into the environment from uses in which the lead was more easily recoverable, such as lead used in auto batteries.

How does industrial ecology fit with other contemporary concepts concerned with improving environmental policy? "Sustainability" and "sustainable development"—meeting the needs of the present without compromising the ability to meet the needs of the future—are the terms that have most come to represent the long-term vision for the environment in many people's minds.[21] Following the 1992 United Nations Earth Summit, sustainability emerged as the international goal for development and environmental conservation. And it is the key concept in the national consensus articulated by the President's Council on Sustainable Development.[22] Sustainability is abstract and thus often fails to help with concrete choices, and it is too vague to help us know when it is being employed indiscriminately or even deceptively. But sustainability has captured the imagination and attention of so many and has proven to be so valuable in redirecting people's thinking, that it is an important focus of next-generation environmental policy. In fact, the authors of the first industrial ecology textbook have linked the two concepts by referring to industrial ecology as the "science of sustainability."[23]

Some other concepts are good short-term guides to action but may be too narrow or too focused on tools themselves to be adequate guides to policy. Risk analysis, for example, forces decisionmakers to acknowledge the interplay of hazard, pathway, and receptor, the trilogy that must be considered whenever we are concerned with potential harm. Moreover, risk analysis demands that we pay special attention to data and facts. Despite its politically charged history in contemporary law and policy, risk analysis is a valuable tool and a key component of next-generation policymaking (see chapter 10).

Still other contemporary concepts highlight aspects of the policy gaps that must be bridged. Reengineering efforts or the ubiquitous focus on "reinvention" reminds us of the need to think about *how* we might improve the mechanics of environmental efforts. Similarly, total quality management (TQM) has focused attention on process refinement, systemic failures, and continuous improvement. Pollution prevention has motivated analysis beyond "end-of-the-pipe," but it can become simply a "front-of-the-pipe" approach to reducing emissions rather than giving due consideration to optimizing all of the "pipes."

Ideas incorporating efficiency such as Stephan Schmidheiny's "eco-efficiency" or Michael Porter's "resource productivity" underscore another important factor—the need not to squander resources, whether natural or financial. These concepts are especially useful at the firm level, perhaps guiding company behavior as the vision of sustainability guides broader societal approaches.

While recognizing the contributions of other concepts, we see important reasons to focus particular attention on industrial ecology. Industrial ecology joins two essential concepts: (1) attention to the natural world as a system (ecology) and (2) attention to the full cycle of human modification of that environment as well as to institutions, the primary instruments of that modification (industry). Industrial ecology has the potential simultaneously to provide immediate guidance for near-term local issues (such as how to achieve cost-effective reprocessing and reuse of discarded materials) and also to help interpret the long-term significance of major natural and economic flows.

As Robert Socolow of Princeton University has described it, industrial ecology also positions corporate entities—from service companies to manufacturing companies to mining corporations to giant agricultural operations—as key players in the protection of the environment. The first-generation view—which sees corporations as reprobates—causes us to miss out on their potential for environmental leadership, especially on the technological aspects of environmental problems. By looking to industry for environmental benefit, industrial ecology also emphasizes that industrial processes and design are important determinants of how resources are and can be used.

Industrial ecology has largely been viewed as descriptive—a science to characterize how the world works, or, in Robert Frosch's language, to describe "an industrial ecology." Whether and how to use the insights gained for policymaking is much less well developed. Robert White, while president of the National Academy of Engineering, offered a definition of industrial ecology that included both its scientific underpinnings and its implications for policy:

> Industrial ecology is the study of the flows of materials and energy in industrial and consumer activities, of the effects of these flows on the environment, and of the influences of economic, political, regulatory, and social factors on the flow, use, and transformation of resources.[24]

In White's threefold focus, systems analysis techniques serve as only the first part of the equation, described as "the flows of materials and energy in industrial and consumer activities." The second part requires additional quantitative and qualitative analysis to measure "the effects of these flows on the environment." Finally, the third part is necessary to make connections to policy—"the influences of economic, political, regulatory, and social factors." Following the example of tracking lead mentioned above, the first part would be the mass-flow analysis, the second part would account for the environmental damage lead creates, and the third part would compel us to ask questions such as: "what if other nations prohibited leaded gasoline as the United States has chosen to do—what would the impact be?"[25] Presumably, now, modeling this policy and its impacts would be within our reach.

In fact, mass-flow analysis can yield significant policy insight. Using this tool, industrial ecologists are finding that not only is the carbon cycle altered by human activity, with the potential for global climate change, but the nitrogen cycle, too, has been disturbed—surprisingly, less by nitrogen oxides emitted into the air than by loadings from the agricultural fertilizers we have been using to feed a growing population.[26] While the precise effects are still being examined, this analysis suggests that we may have to refocus our agricultural policy (see chapter 14).

Part of the "subversive" power for policy of mass-flow analysis, Socolow says, is that "it treats with indifference both what is easy to regulate and what is hard to regulate."[27] Thus, mass-flow analysis identifies where harmful materials are regardless of whether they fall under the regulatory spotlight. Targeting policy attention and inevitably limited environmental resources on the most damaging instances of environmental harm wherever they occur is a critical task for the next-generation environmental protection program. Over time we can modify our regulatory structure to catch up with our scientific knowledge, pushing the boundaries of our scope of analysis, as described in the final section of this chapter.

Another transboundary tool of industrial ecology, life-cycle analysis, yields other policy insights. Life-cycle analysis transcends the fragmentation of the air/water/waste paradigm by tracing the inputs and effluents of all three categories. Using a life-cycle model, analyst John Schall found, for example, that the environmental impacts of recycling municipal waste were no less than the effects of burning or burying it. However, looking at the whole life-cycle—that is, not only at the waste management system but also at the production system that precedes it—creates quite different results. Schall found that recycling is valuable not because it is a superior disposal technique but because the environmental impacts of production using recycled materials are an order of magnitude less than when virgin materials are used in the manufacturing process.[28] Such knowledge provides quite different incentives for producers to become more involved with materials recycling but raises new policy problems because the public sector may be less able to justify recycling as a *disposal* alternative.

In the language of business, a product life-cycle is known as the "value chain," where firms perform value-creating activities from the natural resource beginning to the product end. Value-chain thinking, as David Rejeski shows, can greatly expand the boundaries of environmental learning and management across functions and across firms. When Motorola needed to eliminate

CFC's from its production processes, it turned to its chain of suppliers for assistance in finding substitutes. The capacity for environmental problem solving can translate into interfirm agreements or information sharing and can also include the adoption of environmental management systems and common standards across the value chain, as the International Organization for Standardization is promoting through its ISO 14000 environmental standard setting process.[29]

The fact that both mass-flow analysis and life-cycle analysis are so data-intensive may lead some to conclude that industrial ecology requires an unachievable level of comprehensive knowledge before any policy decision is likely to be made. But these tools of industrial ecology confine themselves to being comprehensive along one axis of a multidimensional equation—tracing a single flow in the case of mass-flow analysis, or the characteristics of an individual product or process in the case of life-cycle analysis. Nonetheless, the effort to understand and connect the industrial and natural worlds is a large and important endeavor that requires us to regularly scan the horizon and fundamentally reexamine whether we have defined our policy problems accurately.

Political economist Charles Lindblom has written that knowledge is partial and policy progress must therefore be incremental. He observes: "Policy is not made once and for all, it is made and re-made endlessly. Policy-making is a process of successive approximations to some desired objectives in which what is desired itself continues to change under reconsideration."[30] Oddly enough, our existing regime of environmental laws makes it difficult to pursue this logic. It locks in standards to the picogram and mandates control technologies almost by brand name. Too often it is inflexible and absolute. Perhaps most fundamentally, it has been an obstacle to what the ecologist calls "adaptive behavior," or the organizational psychologist calls "learning."

A policy model built on industrial ecology, on the other hand, is less likely to get stuck. Because it emphasizes the importance of finding and incorporating new data and practices as our understanding of physical, biological, and political phenomena changes, it produces more flexible and enduring policy inputs. In fact, the tools of industrial ecology are important mechanisms for problem identification, precisely because they are data-driven and fact-friendly. The next generation of environmental policy will depend heavily on our ability to detect and examine emerging phenomena rather than be blindsided by them.

Questions of policy always include not only "what shall we do?" but also "who shall do it?" The institution that faces the largest challenge is government, since it must create the rules, not only live by them. But with industrial ecology the business community becomes, in Socolow's words, "a policymaker, not a policy taker. Industry demonstrates that environmental objectives are no longer alien, to be resisted and then accommodated reluctantly. Rather these objectives are part of the fabric of production, like worker safety and consumer satisfaction."[31]

We must return to the unanswered question of how to find an approach that will overcome some of the cultural barriers to change discussed earlier. The metaphor of sustainability has been evocative. The power of industrial ecology is that it offers a connection to a whole system, not just a fragment. In the dialectic created between human and natural systems, industrial ecology allows us to think past the culture of fragmentation to the specific ends of policymaking.

Managing the Transition

Policy reform is always a difficult process because of the entrenched power of the status quo. The culture of the existing regulatory system with its daunting complexity, as discussed above, has a powerful impact on the thought patterns and the imaginations of all who function within it. Even with its weaknesses, this culture preserves important values that we are not prepared to lose. The fear that reform might lead to less environmental protection is real and must be confronted.

In broad terms, we see three approaches to accomplishing environmental policy reform—"revolutionary," "conservative," and "evolutionary." Some experts advocate policy revolution and press for a complete break with the existing system. Many critics of the initial approach of Congress during 1994–96 saw its proposed reforms as a sweeping deregulatory revolution. Others seek to overcome fragmentation by supporting a single new environmental statute to replace all existing environmental law.[32] But advocates of a statutory overhaul, even one with a serious commitment to environmental protection, quickly find themselves at sword points with policy "conservatives," in this case many environmentalists, who, despite the problems with current laws, are wary of any major change. They see anything more than tinkering on the margins of the existing system as too risky. The irony is that both would-be revolutionaries and conservatives end up paralyzed.

Evolution emerges as the most viable reform approach. Coherent change can best be accomplished through evolving efforts of informed trial and error where the successful new policy species survive and others do not. This approach—perhaps conceived of as nature's way of reform—anticipates mistakes and failures in light of problems of complexity and the rule of unintended consequences. This ecological model implies that fundamental policy restructuring can be accomplished while leaving the current legislative and regulatory system in place. The old gives way to the new only as space and time to test and refine successful reforms permit. Industrial ecology fits with a basic natural metaphor—that we shed the old coat only when the new one is ready.

The proposal to have two competing sets of environmental policies operating in tandem may seem incomprehensible to some observers. Environmental advocates may fear that the dual system will be manipulated by polluters. At the same time, the regulated community is also likely to view dual programs warily, fearing that any new programs will simply add to their regulatory obligations.[33] Both concerns can best be confronted by retaining the existing legal structure, but at the same time consciously starting to replace it with a new system that focuses, through the lens of industrial ecology, on building up new approaches and, ultimately, new policies.

Regulated entities must be encouraged to conceive, propose, and test a diverse set of new practices: the key building blocks of a policy transition. The experiments may be conducted at the facility level but should offer the promise to alter significantly the way in which resources or materials across an industry are used. A *practice*, as discussed here, is a procedure for solving a generic dilemma. It is both an operation and a perspective on how widely it will be effective or on why the operation is likely to be significant in other venues.[34] Xerox, for example, has established an innovative covenant with its customers for lease and buy-back arrangements for photocopying machines designed to facilitate better recycling of its products at the end of their useful lives. Xerox's initial work was very specific; but it is, in the process, establishing a practice about the relationship between equipment producers and their customers that can be employed by others. In fact, it is part of a whole movement in industrial ecology that advocates leasing as a way to assure longer-lived assets.[35]

As a result of many discrete experiments, our stockpile of linked and linkable practices will grow over time. The aggregation of problems solved by the new practices will begin to call into question the way existing law has conceptualized a problem, will encourage the adoption of a new approach, and will ultimately require new policy.

The new and old systems inevitably will clash, since some of the practices developed will conflict with existing law. Waivers from existing rules should be granted for new practices so long as they are more protective than existing alternatives. Of course, this may not be easy to determine. To date, the several EPA programs that provide for tests of alternative approaches have been understandably timid. Thus far, we have little experience with the regular issuance of waivers designed to permit activities that yield a better environmental result than the already existing, legal ones.[36] We still must explore what will be entailed if we actively seek to grant waivers of the scale and scope suggested by adoption of the goals of industrial ecology.

Any evolutionary approach must be guided by its larger goal. It is essential that the alternative regime not be seen as a process to grant site-by-site, facility-by-facility, or instance-by-instance "releases" from normal requirements. Indeed, one criterion for agreeing to consider and monitor a new practice should be that it is a candidate for becoming a routine alternate way of addressing a recurring type of environmental management challenge. Economic actors at every level must be encouraged to develop and implement projects intended to become practices that attain better and less costly environmental results in the spheres in which they operate. Here, the old command-and-control system is both the motivator (because the regulated community will be anxious to escape its oppression) and the equalizer (because the old system will continue to apply to all regulated entities that do not obtain the exemption).

Industrial ecology is not a panacea for environmental policy. Many of the difficulties in environmental policymaking are challenges of governance, knowledge, values, and cost that transcend questions of analytical framework. But with a process of incremental advances building on past advances and on a systems-based understanding of the problems we face, we may be able to create an environmental management system founded in industrial ecology that wins the confidence of policy revolutionaries and conservatives, as well as those in the vast middle ground. As the number of successfully implemented practices grows, they will begin to replace the current system, both informally and formally, through regulation and legislation. Industrial ecology offers an analytic framework for the accumulation of such practices which, when stitched together, can become the fabric of a new environmental policy needed in a world where the interactions between nature and human society daily become more complex. As those practices are given the status of public policy, we can shed our frayed air-water-waste coat and be on a new path to a sustainable America.

Notes

1. See Myron F. Uman, ed., *Keeping Pace with Science and Engineering: Case Studies in Environmental Regulation* (Washington, D.C.: National Academy Press, 1993), a volume from the National Academy of Engineering, for examples of how regulation has not been able to keep up with increases in knowledge.

2. T. E. Graedel and B. R. Allenby, *Industrial Ecology* (Englewood Cliffs, N.J.: Prentice Hall, 1995).

3. The National Environmental Policy Act, which has become a relatively ineffective environmental policy tool, still offers a useful vision of how environmental policy should be made. In particular, it seeks to assure: (1) intergenerational equity ("fulfill the responsibilities for each generation as trustee of the environment for succeeding generations"); (2) environmental justice ("assure for all Americans safe, healthful, productive and esthetically and culturally pleasing surroundings"); (3) beneficial use ("attain the widest range of beneficial uses of the environment without degradation, risk to health or safety, or other undesirable and unintended consequences"); (4) ecological diversity and individual liberty ("preserve important historic, cultural, and natural aspects of our national heritage and maintain, wherever possible, an environment which supports diversity and variety of individual choice"); (5) prosperity ("achieving a balance between population and resource use which will permit high standards of living and wide sharing of life's amenities"); (6) conservation ("enhance the quality of renewable resources and approach the maximum attainable recycling of depletable resources"). See J. McElfish and E. Parker *Rediscovering the National Environmental Policy Act* (Washington, D.C.: Environmental Law Institute, 1995).

4. According to James E. Krier and Mark Brownstein, "On Integrated Pollution Control," *Environmental Law* 22 (1991): 121, whether or not air/water/waste was the best organizing principle for EPA, media-specific bills had been enacted by Congress and the objective of EPA's first administrator, William Ruckleshaus, was to establish the agency as a responsive player. "Ruckleshaus thought that it would be too unsettling, confusing, and time-consuming to begin the new Agency's life with efforts to revamp this fragmented (non)system in favor of an approach organized around administrative functions—such as research, monitoring, standard-setting, enforcement, and the like."

5. Harold D. Lasswell, "From Fragmentation to Configuration," *Policy Sciences* 2 (1971): 439–46.

6. A very useful two-page diagram that illustrates just how different these laws were on issues such as the designation of different federal regulatory agencies, different definitions of effect, approach to risk, and so forth is found in *Neurotoxicity: Identifying and Controlling Poisons of the Nervous System* (U.S. Congress, Office of Technology Assessment, April 1990).

7. One example was inspired by the public disclosure provisions of the "Emergency Planning and Community Right to Know Act." The law has led some firms to dispose of waste by deep well injection rather than find an alternative, environmentally (or economically) preferable means of disposal because waste deposited in that way was not subject to reporting requirements.

8. In particular, CERCLA seems to have created implicit incentives for existing owners to avoid knowing about possible contamination on their property and for prospective users to avoid involvement in cleanup issues related to such property. The unintentional result—until a series of recent initiatives made as part of EPA's Brownfields agenda—may well have been a significant rise in abandonment of urban commercial and industrial properties. Some experts believe that the actual risks to unwitting urban users of such abandoned contaminated properties may exceed the risk averted by the Superfund program. Surely the impact of CERCLA on urban blight is well recognized to be an extraordinary indirect cost of the current Superfund.

9. Philip K. Howard's *The Death of Common Sense: How Law Is Suffocating America* (New York: Random House, 1994) succeeds in major part because of the poignancy of his examples from the environment.

10. See James McElfish, "Back to the Future," *Environmental Forum* 12, no. 5 (1995): 14–23. McElfish, it should be noted, believes that the entire contemporary desire to pursue the next generation of collaborative and holistic environmental policy needs no additional authorization (though perhaps it could use congressional reaffirmation) since NEPA provides what is required.

11. See also Reid Lifset and Charles W. Powers, "Industrial Ecology and the Next Generation Project" (drafted in preparation for the Next Generation workshop on industrial ecology at Yale University, New Haven, March 1996). Lifset, a pioneer of industrial ecology, is now editor of the first peer-reviewed journal to serve this new field, the *Journal of Industrial Ecology* (MIT Press), and he very thoughtfully reviewed this chapter.

12. The first main conference report on the topic of industrial ecology is from L. W. Jelinski et al., *Proceedings of the National Academy of Sciences* 89 (1992), based on a colloquium entitled *Industrial Ecology,* organized by C. K. N. Patel, held in May 1991 at the National Academy of Sciences, Washington, D.C.

13. See Suren Erkman's *Industrial Ecology: A Historical View* (Geneva: Industrial Maturation Multiplier [IMM], 1997), and *Journal of Cleaner Production,* forthcoming, for the strands industrial ecology has drawn together, particularly from the U.S., Europe, and Japan. Many previous analytic tools such as life-cycle costing, energy analysis, and residuals management are antecedent to current methods of life-cycle analysis. The call for papers issued by the new *Journal of Industrial Ecology* states that it "will address a series

of related topics" and then lists material and energy flow studies (industrial metabolism); technological change; dematerialization and decarbonization; life-cycle planning, design and assessment; design for the environment; extended producer responsibility (product stewardship); eco-industrial parks (industrial symbiosis); product-oriented environmental policy; and eco-efficiency.

14. "Life-cycle assessment is an objective process to evaluate the environmental burdens associated with a product, process, or activity by identifying and quantifying energy and material usage and environmental releases, to assess the impact of those energy and material uses and releases on the environment, and to evaluate and implement opportunities to effect environmental improvements. The assessment includes the entire life cycle of the product, process, or activity, encompassing extracting and processing raw materials; manufacturing, transportation, and distribution; use/re-use/maintenance; recycling; and final disposal." Society of Environmental Toxicology and Chemistry, *A Technical Framework for Life-Cycle Assessment* (Washington, D.C.: SETAC and SETAC Foundation for Environmental Education, Inc., January 1991), chap. 10.

15. See Battelle, Pacific Northwest Laboratory, "The Source of Value: An Executive Briefing and Sourcebook on Industrial Ecology" (February 1996), 3.2.

16. Quoted in Ernest Lowe and John Warren, *The Source of Value: An Executive Briefing and Sourcebook on Industrial Ecology* (Richland, Wash.: Pacific Northwest Laboratory, 1996), 3.11.

17. Nicholas Gertler and John Ehrenfeld, "A Down to Earth Approach to Clean Production," *Technology Review*, February–March 1996.

18. Robert Frosch, "Industrial Ecology: A Philosophical Introduction," *Proceedings of the National Academy of Sciences* 89, no. 3 (1992).

19. The systematic tracing of materials and energy flows from extraction of materials from the earth through industrial and consumer systems to the final disposal of wastes was named "industrial metabolism" by Robert Ayres, its founder. See, for example, Robert Ayres, "Industrial Metabolism," in *Technology and Environment*, ed. Jesse H. Ausubul and Hedy E. Sladovich (Washington, D.C.: National Academy Press, 1989).

20. Valerie Thomas and Thomas Spiro, "Emissions and Exposure to Metals: Cadmium and Lead," *Industrial Ecology and Global Change*, ed. Robert Socolow et al. (Cambridge: Cambridge University Press, 1994).

21. The most frequently referred to source on this is World Commission on Environment and Development, *Our Common Future* (Oxford and New York: Oxford University Press, 1987).

22. President's Council on Sustainable Development (PCSD), *Sustainable America: A New Consensus for the Future* (Washington, D.C., February 1996).

23. Graedel and Allenby, *Industrial Ecology*.

24. Robert White, preface to Allenby and Richards, *Greening*.

25. Robert Socolow and Valerie Thomas, "The Industrial Ecology of Lead and Electric Vehicles," *Journal of Industrial Ecology* 1, no. 1 (1997).

26. Robert Ayres, William Schlesinger, and Robert Socolow, "Human Impacts on

the Carbon and Nitrogen Cycles," in Socolow et al., *Industrial Ecology and Global Change,* 121–55.

27. Robert Socolow, "Six Perspectives from Industrial Ecology," in Socolow et al., *Industrial Ecology and Global Change,* 3–16.

28. John Schall, "Does the Solid Waste Management Hierarchy Make Sense?" Program on Solid Waste Policy Working Paper no. 1 (New Haven: Yale University Program on Solid Waste Policy, 1992).

29. David Rejeski, "Clean Production and the Post Command-and-Control Paradigm," in *Environmental Management Systems and Cleaner Production* (forthcoming).

30. Charles Lindblom, "The Science of 'Muddling Through,'" *Public Administration Review* 19 (1959): 79.

31. See NEPI, Reinventing EPA and Environmental Policy Working Group, the Unified Statute Sector, "Integrating Environmental Policy: A Blueprint for 21st Century Environmentalism" (Washington, D.C.: NEPI, 1996).

32. Socolow, "Six Perspectives," 12–13.

33. See, for example, Frederick Anderson, "From Voluntary to Regulatory Pollution Prevention," in Allenby and Richards, *Greening,* 98–107. Anderson concludes that the regulated community should limit implementation of programs not required by regulation—irrespective of their salutary effect on the environment—to situations where the programs can be justified solely on economic grounds, and that they should carefully weigh those benefits against the likelihood that they will generate regulatory experience that enables rapid deployment of a second and unprecedentedly "intrusive regulatory system" for pollution prevention which will simply be cobbled together with the existing system.

34. Prof. Tim W. Clark, Yale University, personal communication, 1997. See also Ronald D. Brunner and Tim W. Clark, "A Practice-based Approach to Ecosystem Management," *Conservation Biology* 10, no. 5 (October 1996): 1–12, which describes the need for the building and testing of practices at the level of the ecosystem in a parallel way to the need described here for new practices compatible with industrial ecology. Brunner and Clark (p. 2) explain the need to be evolutionary because "ecosystem contexts are far too diverse, complex, and dynamic for anyone to understand completely, completely objectively, and once and for all."

35. See, for example, Lowe and Warren, "Product Life-Extension and the Service Economy," *Source of Value,* chap. 4.

36. The *Sustainable America* report did stress that site-specific waiver programs would require far more regulator time and effort (and authority and discretion, incidentally) than would normal compliance and enforcement efforts.

t w o

Ecosystem Management and Economic Development

John Gordon and Jane Coppock

Protecting nature and developing the economy have often been viewed as separate, if not opposing, activities. Frequently it has seemed to come down to an either/or choice: either seal off an area from economic development to protect threatened species or support the economy by putting as few environmental constraints as possible on new projects. Environmental regulation has been seen as the creation of a central government that showed little or no understanding of its effect on people's livelihoods. Economic development has been perceived as being carried out with little or no regard for the damage it was doing to vital or irreplaceable natural systems. These views have lent themselves to caricature as "jobs versus the environment," or "loggers versus spotted owls," or "dams versus snail darters," and in extreme cases have resulted in heated conflict. Economic and environmental opportunities have been lost. Money and time have been wasted in endless court battles.

A new, less fractious, more collaborative approach to settling these disputes is clearly needed and has been developing in pockets around the United States. This new approach takes seriously the need both to protect habitat and to promote economic growth. It looks for ways to simultaneously achieve protection and development using policy tools to connect, not separate, them. An integrated policy approach represents a significant break with the past, where "pristine" environments and economic growth maximization were both held to be sacred, and attempts to bring them closer together were viewed as threatening. Discovering how to avoid the cost of separating environment and economy without endangering either value is at the heart of the search for new approaches, and is an important foundation for the next generation of policy.

Some of the ferocity of the battles over environment and development has resulted from a lack of appropriate concepts and practical tools to analyze the circumstances scientifically and to handle them politically and socially. Ecosystem management is both a concept and a tool that makes the extensive developments in ecological science applicable and usable for people in the field. New methods of dispute resolution have advanced the process of collective decision-making. Without these tools, people insist on prohibition and separation as the only means of protection on both sides. With these tools it becomes possible to create a common knowledge base, encourage a comprehensive perspective, and establish the dialogue necessary to begin to build trust.

What is the larger goal? With regard to both human and natural systems, "caring for the present without destroying the future" seems to capture the essence of what many people value. Increasingly in the United States, we must try to achieve that goal within a finite landscape and with a growing human population. In previous eras, conflicts over the same piece of territory could be resolved by going elsewhere, to a new patch of ground. We solved our conflicts by doling out land to whomever wanted it or appeared to need it. We must now choose how much of our land to occupy, to develop, or to preserve in order to have both the economic opportunity and the environmental quality that the vast majority of citizens want. Often, the same tract of land must accommodate both preservation and development.

The shift in attitude we are starting to see toward environment and development may represent the tangible realization that the United States has begun to resemble other, more crowded countries. Increasing competition for scarce environmental and economic resources means that it is unlikely that constituencies will get everything they want. In light of this very different assumption, the next generation of environmental policy must use new tools to identify and focus on the still plentiful circumstances in which environment and economy can coexist profitably.

Ecosystem Management

One important vehicle for achieving greater integration is a broad, systems-based approach that looks at the overall structure and behavior of a given

Complex mathematical and diagramatic treatments of the principles of ecosystem management (EM) could and should be assembled and, to some degree, have been. The minimum set of EM instructions, however, includes five simply stated concepts. Their simplicity of statement is misleading because they are easy to say but very hard to carry out.

Manage Where You Are
Because EM requires both the recognition and the transcendence of real boundaries, it applies to a specific site. Although the notion of *site specificity* is old in forestry, the realization of EM will require even greater emphasis on the specific properties of the plan to be managed and the objectives for which it is managed. Many of the examples of "poor" past practice are results of moving "best practices" out of their effective domain defined by places and objectives.

Manage with People in Mind
All earth ecosystems are subject to human influence, known and unknown, overt and accidental. Management itself implies strong human purpose. Indeed, some of the most difficult tasks proposed for EM revolve around the notion of excluding human use and influence from specific systems. Thus, EM begins with a careful evaluation of human desires, influences, and responsibilities. Every system definable in biological and physical terms connects to and interacts with a network of human values, uses, institutions, and other social structures.

Manage Across Boundaries
EM tracks transactions across defined system boundaries and moves the boundaries themselves when necessary. At the most elementary level, this requires that "neighbor" influences be recognized and managed. This often must be done without extending fee ownership and through a complex process of joint goal setting, compromise, regulation, and incentives.

Manage Based on Mechanisms Rather Than "Rules of Thumb"
Knowledge of the specific processes and interactions responsible for system activities and outputs is the key to continuously improving EM. Landscape-based systems are so complex that they can rarely be managed by enumeration and tracking of all components. Detailed knowledge of why things occur as they do is thus the only path to the prediction of system behavior and output. This requires reliance on sampling and on knowledge gained elsewhere but tested and adapted in the system at hand.

Manage Without Externalities
In the sense that it is used here, *externalities* means system contents not currently seen to be related to management objectives. At first glance this seems to contradict the assertion that complete enumeration of system contents is seldom, if ever, achieved. However, it requires that all known contents of the system be included and considered when decisions and manipulations are made.

2.1 Simple Ecosystem Management Principles

area, such as a watershed or an estuary or even a city, and attempts to give direction to decisionmakers about which components are crucial for the healthy ecological functioning of the area, which components or portions could sustain development, and which components cannot be made to positively fit together, or overlap, or both. This broad approach is often called ecosystem management because it identifies the boundaries of distinct ecological areas ("ecosystems") and is concerned not just with analyzing them, which is the job of the scientist, but also with managing them (fig. 2.1).[1]

The promise of ecosystem management is that it will both produce more environmental protection for whole habitats without having to shut people out completely, and enable more economic development in places that can sustain it. In many places, human interaction can improve rather than degrade the ecology of an area, through, for example, habitat restoration, creation, or maintenance.

In the past, land management has often been focused on general approaches to be applied wherever roughly similar conditions were encountered. Outputs perceived to be beneficial, such as the preservation of a species and the production of a defined quantity of a commodity such as water or wood, were set as goals to be pursued over undefined time frames without further elaboration. Ecosystem management attempts to cure this vagueness by setting boundaries in time and space, understanding the whole system within those boundaries, enumerating "neighbor" influences, and building in learning routines that continually increase insight and predictive ability. Most important, it regards the system not as a bundle of potential outputs but as an evolving whole to be understood and managed contextually, in its own specific place and time.

The core of all ecosystem management projects is research and information. The approach requires a solid science base that has been made possible through a number of recently developed research and information management tools. One of these is geographic information systems (GIS), which allows designers, planners, and researchers to translate many kinds of data—physical, biological, demographic, and economic—into layered maps that reveal patterns and processes in whole geographic regions that cannot be perceived on the ground. For example, watershed-level planning made possible through the use of GIS enables developers to pinpoint where they can build without endangering water quality or where they shouldn't build because of the long-term economic risks of occupying an unstable or flood-prone area.

Computational advances that make possible the creation of large-scale models and the structuring of large data sets allow the consideration of much greater ranges of management variables than even a few years ago. Ever more mobile and precise monitoring equipment permits tracking in the field of a larger number of variables with greater accuracy.

The major scientific advances that underlie the promise of ecosystem management are derived from ecology. The science of ecology is a potentially strong foundation for ecosystem management, as well as an impediment to it.

Ecological science, as currently practiced, is both at the cutting edge of understanding complex systems and at a crisis point in its development. Much of the past effort to understand the relationships of organisms to their environment, individually and in aggregate, has been focused on "pristine" areas where human influence is low or nil. The assumption was that people were an aberrant influence antithetical to understanding nature. But ecosystem management exists to serve people in nature, and the most critical and complicated management questions arise from human influence. The study of ecology must increasingly include the study of humans as ecological factors if ecosystem management is to succeed.

In another dimension, ecological observation and experimentation have rarely been carried out in ways conducive to making the results "add up" as a management tool. The eminent ecologist Harold Mooney characterizes this as "Frank Sinatra science"—everyone does it "my way." Experiments are rarely linked across geography and subdisciplines by common experimental designs or methods, research activities have often been restricted to small plots and short times, and management activities, which take place over larger areas and longer times, have not been linked to research. The cultures and reward systems of researchers and managers are vastly different. Until these cultures are made more compatible, ecosystem management will not reach its potential. Scientists must learn more about management, and care more about it. Managers must learn more about science and particularly about scientists, and be able to include science in their management agenda. Both must move toward adaptive management that makes learning and the use of specific knowledge to guide future decisions a primary output.

Blending Policy and Process

What we see emerging in individual settings around the country is new skill at applying ecosystem management principles and handling the politics of collaboration. In these instances the potential for competing interests to work toward solutions is increased because it is possible to be detailed and specific about how the integration might actually work. Fort Ord, a twenty-eight-thousand-acre decommissioned U.S. Army base in Monterey, California, has now been divided into protected areas for endangered species, corridors for light development, nonsensitive areas for commercial development, a new state university, and a research park—all based on a comprehensive habitat conservation plan. The plan designates three primary conservation areas connected by corridors where some economic development is being allowed. Outside the protected zones, the plan permits largely unimpeded development to support economic recovery of the region.

Officials from the state university, the federal government, and the local community are cooperating in the development of a training program for former base employees. Cooperating groups include the Bureau of Land Management (BLM), the U.S. Fish and Wildlife Service, two community resident groups, the Monterey County Board of Supervisors, the University of California, California state park officials, the city of Marina, and the U.S. Army. The Fort Ord project demonstrates both the systems thinking and the cooperative processes that undergird successful ecosystem management efforts.[2]

The potential for conflict between economic and environmental interests takes many forms. But across this spectrum, ecosystem management helps focus the relevant parties on ways to achieve the coexistence of human and natural systems. In the Adirondack Mountains of New York State, environmental protection gave the economy a boost by enabling a natural resource-based industry—fishing—to reemerge. Effluent from the area's lumber and tanning businesses had long caused serious deterioration of the water quality of the Beaver Kill and Willowemoc watersheds. When those industries declined and disappeared, the area was left with few jobs and badly damaged streams. In 1994, Trout Unlimited, a nongovernmental organization, joined state and local officials in an assessment of trout fishing in the watershed. When it was determined that recreational fishing could become a cornerstone of the local economy, Trout Unlimited and the state joined forces and embarked on a long-term plan to manage and conserve the watershed, helping to resuscitate the environment and the economy.

In another instance, shipping—an industry central to the economy of the Great Lakes region—was creating an unacceptable ecological problem. Dumping of ballast water by commercial ships had introduced so many exotic species of aquatic plants and animals that the ecology of the lakes was disturbed. The invasion of foreign fish, crabs, mollusks, bacteria, and viruses, which reproduce explosively, was damaging not only to the shipping industry but also to fisheries and industrial and public water supply systems. Rather than lose this industry, stakeholders came together and ultimately determined that technical solutions could be found to remedy the environmental harm. Seven Great Lakes states jointly agreed to fund a project to conduct new ship technology tests on vessels supplied by a commercial shipper. Project participants brought expertise from the shipping industry, state governments, Canadian and U.S. agencies, nongovernmental organizations, and private engineering firms.[3]

Ecological techniques have also been used to improve long-term prospects for ranchers at a reasonable short-term cost. In Trout Creek, Oregon, destruction of habitat near streams by grazing cattle had led to a reduction of water quality and imminent loss of a pure strain of endangered native trout. The Trout Creek Working Group was formed by ranchers, state, and BLM officials and environmental groups to try to formulate their own plan instead of heading into what looked like an inevitable court battle. It took many, many months of talking for the groups to get past their suspicion and disapproval of each other. What emerged, however, was an understanding of their common commitment to the land, and a management plan that they were willing to implement. The plan involved reducing the size of the herd and pulling cattle away from all of the riparian areas (strips of land bordering streams) to allow the protective vegetation to regenerate. The plan's successful results—better water quality, viability of the endangered fish population, and a sustainable level of ranching—were achieved relatively quickly (three years) and relatively inexpensively (compared to potential court costs or the loss of livelihood altogether).

In all of these cases, it was collaboration among groups with a common interest in the situation at hand that produced a new and more integrated solution. Some of the collaborators were long-time enemies and began the collaborative process with deep distrust. At the Seventh American Forest Congress, held in February 1996 in Washington, D.C., fifteen hundred people sat together for four days to work out a consensus about how they thought American forests should be managed. Working in groups of ten people, including representatives of industry, government, environmental groups, and the general public, the participants attested that they did more listening and made more progress toward consensus than had been achieved through any other means. The few attempts made at confrontational, partisan gestures were largely ignored. The congress participants are now working in their home locations to translate their vision into specific projects.

Adopting Systems-Based Approaches

Examples from around the country show that groups can work positively and set a high environmental standard for what is sustainable over time. This translates into trying to match the given need for employment with the desire for environmental protection in a particular place. It enables the role of regulation to shift from prohibitor to goal setter, providing a broad framework within which context-specific solutions can be developed that benefit the circumstances and limit compromising the future as much as possible for either the environment or the economy.

There are cases at the extreme in which an integrated solution will not do, however, either because the political climate has deteriorated beyond cooperation or because the situation warrants unqualified support on either the economic or environmental side. If more integrated policy becomes the norm, however, the extreme cases should decrease in number. The new policy will instead reflect the desires of partners who want to achieve their goals through information-based collaboration, not win-or-lose confrontation.

More cooperative ecosystem-based processes should also yield more durable arrangements. They involve a broader array of people than past solutions have, and they build on a more comprehensive picture of the resources at stake and the competing interests that must be accommodated. The inclusiveness of the process broadens the base of support, making it harder for die-hard opponents to overturn agreements as soon as they see a political advantage. The new approaches also reflect a wider view of economics. They attempt to bring the benefits of market thinking and entrepreneurial attitudes to the table along with the recognition of the environmental costs, present and future, of any individual development project. At the same time, the process acknowledges the economic and environmental value of traditional nonmarket goods like fresh air, natural vistas, cultural heritage, uncontaminated soil, and water fit for swimming and fishing.

Once this systems perspective with its broader view of costs and benefits is brought into play, opportunities for a different kind of entrepreneurial effort start to appear, an effort that requires the collaboration of partners with very different kinds of expertise, including hydrologists and landscape ecologists, wildlife biologists and experts with GIS training, private developers and government officials. The process represents a healthy democratization of environmental decisionmaking, incorporating many of the political values emerging in the other policy areas described throughout this book: an enhanced role for local or regional decisionmakers, consciousness of cost/benefit ratios, incentive-driven rather than control-driven regulatory schemes, and working toward a communal or overarching goal in ways that are particularly appropriate for the project at hand.

Despite all of these benefits, however, there are difficulties and pitfalls to be wary of in adopting a more integrated, systems-based approach. Ecosystem management is controversial because it requires us to manage across boundaries, between states, and across private-public property lines. It is controversial because it tries to encompass whole systems, and both ecological and economic systems are dynamic, complex, and hard to predict. Moreover, ecosystem management requires a lot of local participation to work. If the people who live in an area are not happy with the way it is being managed, the management plan will not succeed.

The educational task entailed by decentralized decisionmaking is daunting. Having seen how difficult it has been to inject good science into decisionmaking at the federal level, contemplate the now hundreds of circumstances in which no one has access to detailed scientific information about the specific area. Although local personal knowledge of individual contexts is invaluable and sometimes very rich, it may be insufficient to inform broader management decisions. Therefore, we must often look for appropriate contributions to the management process from the federal, state, and local levels. Extension services, participation of nongovernmental organizations' technical staff, and access to the Internet can further help bridge knowledge gaps.

Most ecosystem management will be practiced on a mixture of public and private ownerships, and will be driven by public and private investment. New and more flexible approaches to financing large-scale projects will be needed, as will new incentives to motivate private participation. Conservation easements, public land acquisition programs, and tax incentives for the maintenance of land types from farms to forests provide a starting point, but much remains to be done. Firms that hold and manage forests as investment opportunities for large private investors, for example, have great potential for the implementation of ecosystem management because of their long time horizons and their needs for predictability and public acceptance.

Even when we accept a changed role for the federal government and charge it with setting goals, allowing smaller units—regions, states, and localities—to decide how to reach the goals, we do not avoid risk. One danger is that the small units will lose sight of their responsibility to balance ecology and economy. By being so flexible, we may lose accountability. The rule of law and the fear of punishment have done much to advance environmental progress, despite all the drawbacks we know that arrangement to have. The rule of law at the highest level must not be lost sight of—there must be stiff penalties for clear harms and no ambiguity about it.

Another danger of the decentralization process that ecosystem management normally entails is that natural resources might not be managed for the good of the nation as a whole. There are mountain ranges and prairies and river basins that belong to all of us and to our descendants, as well as to the local community. Who decides what will happen on or to them? One approach is to say that community is the place decisions should start but that there are two kinds of community—community of place and community of interest. A balance needs to be struck between them to guard against the dangers of short- or narrow-sightedness.

Involving local people in the environmental management of their area does not mean excluding those from the larger community of interest. In some cases, this means including federal agencies to represent the nation's commitment to protected areas. Yet local people still must be involved in decisions that affect their lives and their livelihoods.

Reaching a decision when there is no consensus is a difficult political process. Especially important is acknowledgment of cross-sectoral responsibilities between the public and private realms. Private sector actors need to recognize more fully the limits to ecological resilience their activities sometimes demand. Public sector actors need to recognize the fragility of wealth creation itself—namely, just how nonautomatic it is that businesses succeed and communities thrive. Collaborative decisionmaking is also time-consuming, which creates a temptation to opt for the preemptive move that does not involve the community. What is emerging, however, is the realization that the cost of that form of decisionmaking will end up being higher than the cost of the more cumbersome, slower approach. Skill in dispute resolution and conflict management, however, is essential. In a democratized policy universe, process becomes very important. The better we understand what kinds of socio-political processes work best to support collaborative environmental decisionmaking, the faster we will be able to move beyond confrontational politics.

Forging Ahead

In broad terms, then, a shift in policy toward environment/development integration allows tradeoffs, encourages experimentation in technology, and is outcome-based. It promotes adaptive response to experience and discourages inefficiency, especially in the form of turf wars between competing and overlapping agencies. It focuses on measurable results and creates incentives to achieve those results. It tries to minimize perverse effects, such as destroying habitat because it might contain endangered species. It weighs in heavily against subsidies and tariffs, preferring to expose the true costs and benefits of integrated economic development, environmental protection, and government regulation.

The limits to reconciling economic development and environmental management must also be recognized. It is hard to reconcile competing interests when resources or other opportunities are very restricted. If a species is three individuals away from extinction, there is not a lot of latitude. If people are so polarized that there is no chance for broad involvement, a systems approach that requires longer-term planning won't work. When people have decided they would rather fight than win, there is not much room for either political or technological solutions.

From the perspective of process, the greatest enemy of the next generation of solutions is the notion that you either have consensus or you have a dog fight. In a democracy, achieving a consensus that makes everyone happy is rare. Instead, we have to operate with the broad array of people in the middle who will agree that there is a workable solution even though it is not optimal. The true believers will defend the environment or economic development against all opposition. We are never going to bring them along. We are going to have to learn to operate with support from alliances that represent a preponderance of public opinion, but certainly not a consensus.

Magical transformations of attitudes or policies are not likely. But ecosystem management offers a way to shift the terms of the debate, a redirection that reflects greater knowledge of natural systems, greater skill at devising blended solutions to conflicts, and the prospect of both environmental protection and sustainable economic development.

three

Property Rights and Responsibilities

Carol M. Rose

Environmental protection often seems to be at loggerheads with private property. In the United States, as in other parts of the world, the claim is often heard that environmental regulation deprives private property owners of part (or all) of the value of their land. Much of the concern arises from the burdens placed on particular individuals, who may find, for example, that they are being asked to preserve wetlands or endangered species habitat on land they had planned to develop. Such individuals claim that they are being unfairly singled out to bear the costs of public environmental protection programs.

In fact, both secure property rights and effective environmental protection share a common goal—the enhancement of the total social well-being, both private and public. Frictions often arise when new understandings about environmental harms confront preexisting property interests and settled landowner expectations. But in general, the safeguarding of property, far from conflicting with environmental protection, can be an extremely important vehicle to smooth those frictions, and indeed to carry out a fair, frugal, and effective program of modern environmental regulation.

Public and Private Property Rights

Private property rights are essential in a free-enterprise regime.[1] An owner must have reasonably secure expectations of continued ownership if he or she is going to expend efforts to improve resources. Similarly,

* Some portions of this chapter are taken from Carol M. Rose, "A Dozen Propositions on Private Property, Public Rights, and the New Takings Legislation," *Washington and Lee Law Review* 53 (1996): 265.

reasonably secure definitions of property are essential to trade, since trading partners must know who has what in order for their trades to mean anything. These elementary building blocks of capitalism—encouragement to labor and trade—are important reasons for making property secure, and they are very widely recognized in the common law of property.

But although property rights need to be reasonably secure, their content has always changed with changing social and economic conditions. Property rights in traditional law have never had fixed characteristics that apply under all conditions and for all time. Indeed, it would be undesirable and probably impossible for property rights to have such fixed definitions.[2] Why do property rights change over time? One chief reason is that it is costly to establish property rights, and hence there is no point in doing so until the need arises. History tells us that in most instances, people only begin to define property rights when resources become scarce. This is in fact a typical pattern in common law property rights. For example, grazing rights were only very loosely defined in the early years of settlement in the West, but they became much more sharply defined as more settlers arrived with more grazing animals, which raised the possibilities for strife over grasslands.[3]

This pattern responds to the benefits and costs of establishing and defining property rights. There is nothing wrong with unrestricted, open-access common usage when a resource is sufficiently plentiful. But when users and uses increase, congestion increases too, and an open-access resource may begin to deteriorate. This situation is often called "the tragedy of the commons," and the tragedy is especially likely to strike at environmental resources like air, water, and wildlife. The "tragedy" is that everyone's individual incentive is to keep on using the resource to the maximum, when in fact unrestrained individual use will exhaust the resource as a whole, to the great detriment of all the users.

One way to restrain individual uses, and to avoid congestion and strife over open-access resources, is to divide the open-access "commons" into private property. Private property of course has the advantage of encouraging people to be careful about the resources under their control. On the other hand, some resources, because of their scale and complexity, may benefit from larger-scale management—even public management—in order to safeguard what have traditionally been called "public rights."[4] The use of waterways, for example, has been considered a public property right more or less continually since Roman times.

The waterway example illustrates another very important feature of property rights: it is easier to define private property rights in some resources than in others. Land is fixed in location and can be visibly marked with fences or other boundary markers, and trespassers can be relatively easily identified. But the boundaries of water rights are considerably more difficult to demarcate. Stocks of wild animals and fish are similar to water in that they move around and cannot easily be designated as belonging to one person or another. Most difficult of all to "propertize" is air. The difficulty of defining and enforcing private property rights in air, water, and wildlife has never meant that these environmental resources are not valuable, but simply that they may be more easily managed as *public* or *common* property.

When people acquire individual property rights in land, they often enjoy access to adjacent environmental resources. That is, they effectively "piggyback" the use of these common resources onto their private land. For example, a landowner may fish in a nearby river or use the air to disperse smoke and soot. Such uses are not a problem so long as the air, water, or other common resources are relatively plentiful. As with individual property rights, there is no particular need to assert and formalize public rights in common resources so long as these resources remain plentiful.

But where population grows more dense, these unrestricted uses of the commons can become a problem. That is why London had restrictions on burning coal as long ago as the thirteenth century. That is why early nineteenth-century American states restricted access to shellfish in their tideland waters. That is why later nineteenth-century American law increasingly recognized rights of action for nuisance against landowners who caused undue smoke, fumes, noise, and water pollution.

Simply put, even long-standing use of common resources does not create a *right* to future use. As circumstances change and what might once have been seen as a "normal" activity is recognized to limit the opportunities of others to use the same resource, public rights will be, and for efficiency's sake must be, reasserted.[5]

The Takings Controversy

In the recent controversies over property rights in environmental policymaking, some have asserted that the Constitution's "takings clause" requires property owners to be compensated whenever regulation reduces the value of their land. Though some of these claims are extreme, compensation to private owners has unquestionably formed a part of traditional Anglo-American law. Certainly courts and legislatures have generally recognized a duty to compensate owners for holdings that are appropriated outright for public uses.

But historically, any duty to compensate was subject to several countervailing principles. Most notably, compensation was generally not thought due where large numbers of landowners shared more or less equal regulatory burdens, since such scenarios might be likened to a type of in-kind tax, not singling out particular owners disproportionately. If, for example, all the property owners in a town are required to clear the snow from the sidewalks in front of their houses, the burden of snow removal is roughly evenly distributed, although some people will have to do somewhat more than others. Nor was compensation due when regulatory burdens were minor, or when they were implicitly recompensed by reciprocal benefits or "set-offs" going to the affected landowners.

Moreover, compensation was not due when regulation simply prevented private owners from doing something to which they were not entitled. Zoning restrictions, for instance, may prevent landowners from setting up factories in residential neighborhoods, but they have long been upheld as a prevention of nuisancelike damage to the surrounding owners. And more generally, private owners were not entitled to damage public rights in environmental resources, because damage to such resources was itself seen as an assault on other peoples' property, however diffuse the "owners" might be. Dumping garbage in harbors or debris in running streams, for example, has long been forbidden.[6] Traditional American law has never regarded landownership as a license for the indefinite or unrestricted use of adjacent resources such as water, air, and wildlife, particularly in situations in which one landowner's use could have serious or cumulative effects on others.[7]

Nineteenth-century legislatures sometimes explicitly *authorized* industrial or development activities (such as factories and railroads) that might otherwise have been subject to abatement as public nuisances because of the air or water pollution and wildlife losses they caused. The original concept behind permitting such inroads on public rights, however, was one that is still instructive today: the theory was that any damage to public rights had to be justified by an even greater benefit to the public's well-being. Indeed, through a revived "public trust" doctrine and an increasingly sophisticated analysis of private rights, later nineteenth-century courts kept tabs on legislatures, to make certain that authorizations of public nuisances were not simply disguised give-aways of public rights to privileged private individuals or firms. In rapidly developing New York, for example, the nineteenth-century courts scrutinized railway franchises increasingly closely, to make sure that neighboring owners would not be subjected to undue pollution, noise, and disruption; and the U.S. Supreme Court itself used a public trust doctrine to invalidate a legislative give-away of extensive public waterfront to private development.[8]

But the increasingly frequent pattern for the courts was to defer to the judgment of legislatures and public prosecutors in the protection of public rights. Effectively, the courts recognized the need not only for change but also for the more sophisticated and systematic management of common resources that legislation could provide. As population increased and knowledge about pollution grew, courts in the later nineteenth century recognized a wider scope for more complex and comprehensive regulation to protect air, water, and wildlife.[9] We are much more aware today of the impact of human uses on common environmental resources, but modern environmental laws are the successors to thirteenth-century London's prohibitions on coal burning, the early American restrictions on obstructions to waterways, the later nineteenth-century public assertion of responsibility for protecting fish and wildlife stocks, and a whole panoply of public efforts to protect health, safety and welfare from common resource overuse that was piggybacked onto private property.

Thus the public rights tradition in American law concerned widely used resources that were valuable but costly to privatize. It would have been wasteful—another tragedy of the commons—to allow individual owners to appropriate resources that were effectively shared by many others, and traditional American law did no such thing. Indeed, in protecting environmental resources, our legal institutions also protected property rights, both public and private. But tension remains over the line between private and public property rights, most visibly where takings claims are made.

Transitions and the Need for Compromise

Many aspects of traditional American law encouraged private property owners to be generous in allowing neighbors and others to use their land—for example, for hunting. But the quid pro quo was that the landowners could change their minds when they wanted to control access to the land more intensively, and the neighbors who had until then used the land acquired no permanent rights to continue indefinitely.[10] The same idea applied to *private* uses of *public* resources. The 1915 case *Hadacheck v. Sebastian* (239 U.S. 394) illustrated the point: there it was held that even though a private brickyard could emit smoke and fumes so long as the surrounding areas were lightly populated, public authorities could halt the use when the area became more heavily populated, and when the public was actually more threatened by the noise and air pollution that were effectively private encroachments on public rights.

There were many cases like this in nineteenth-century American law. Thus in traditional American takings law, even if a private landowner had piggybacked the use of public resources like air or water, that fact did not yield permanent rights to the private owner. That is, past private usage of public resources was not necessarily an impediment to legislation that would protect public resources in the future. Indeed, the courts sometimes upheld legislation that effectively dismantled previously existing uses, as in *Hadacheck*.

From a practical perspective, however, such draconian restrictions are not always a good idea, because there are important reasons to be careful in regulating existing uses. One of the unfortunate characteristics of environmental problems is that they are often not visible until a critical threshold is passed.[11] By the time a problem is generally recognized, private owners may have innocently sunk resources into their land uses, perhaps at a time when the damage to adjacent common resources did not in fact amount to much and when the owners simply assumed that they could indefinitely continue to use public resources like air, water, or wildlife stocks. Halting such uses at a later point may result in a loss of those sunk expenditures. Moreover, forcing landowners to discontinue such uses may seem especially harsh if the early uses, taken alone, did not cause great harm,[12] and if the public has been tardy in responding to the damage caused by accumulated private actions or land uses.

Equitable considerations may therefore sometimes argue for exemptions to owners who are particularly caught short by regulatory change, or for giving such owners positive inducements to cease their inroads into public resources rather than flatly prohibiting their uses. That can be the case even though the preservation of the public resources themselves, for all users present and future, argues for limiting any *further* private inroads on public rights.

It is precisely the province of takings law to balance transition problems of this sort—that is, the shift to more stringent but legitimate protections of public resources, under circumstances where private resources have already been sunk into private property on the basis of prior and more permissive regulatory regimes. The issue is not that the prior permissive regime was necessarily an unfair give-away of public resources, since there may have been less pressure on public resources at the earlier time; nor is the issue that the new regulatory regime is inappropriate or unwise, since resource use may now be more intensive. The issue, rather, is that certain private owners may get caught in the transition.

As a practical matter, for those concerned about environmental questions, it is extremely important that these transition problems be solved smoothly and without undue hardship to particular property owners. Quite aside from issues of fairness, in a democracy property owners who feel themselves mistreated may organize to halt the measures that aggrieve them, creating significant opposition to environmental programs. In that sense, an aggressive effort to tighten regulations can backfire unless it includes an appropriate transition strategy.[13]

Given this possibility, both legislatures and courts have created a variety of compromises to avoid serious unfairness and undue burdens to individual property owners, while at the same time preserving the ability of legislatures to protect public rights. For example, new legislation may "grandfather" preexisting uses,[14] though it is of course important that such temporary exemptions be understood to be time-bound, lest they harden into new claims. On their part, courts may require that private property owners with "vested rights" or "investment-backed expectations" be indemnified to avoid having innocent property owners suffer particularly acute losses, even if the legislation in question is otherwise a reasonable effort to protect common resources.

Takings cases represent judicial action to effect such protections of owners' settled expectations from serious and unequal disruptions. In that sense, takings cases can relieve the demoralization of these owners, disarm their potential resentment against environmental law, reassure citizens that they will not be singled out to pay for public programs, and to some degree prevent the social waste of preexisting capital investments made in good faith. But viewed as a whole, takings case law normally represents a compromise. The flip side of that compromise is to permit regulatory bodies, over time, to adjust the protections necessary for the preservation of public rights and resources, without the need for compensation or special dispensations beyond a point at which owners should reasonably adjust their own expectations to a new regulatory regime.

From an environmental policy point of view, accommodation with the settled expectations of property owners would often be better approached legislatively. Efforts to ease the transition to new regulatory regimes might entail:

- phasing out or restructuring counterproductive subsidies rather than restricting private uses;[15]
- permitting less costly but effective "substitute" performance;
- avoiding flat prohibitions and instead allocating limited use rights through taxes, charges, or systems for taking turns;
- phasing in new regulation, permitting relatively undamaging "nonconformities" to continue during specified transitional periods;
- offering partial or temporary exemptions or variances for small owners or for other nonconforming uses that present special hardships;
- providing funds to alleviate special hardship cases;
- making use of administrative boards to consider special circumstances warranting substitute compliance, limited exemptions, or claims for assistance;
- encouraging informal and rapid dispute resolution methods in order to settle property owners' claims promptly and without undue administrative expense;
- supporting local or regional efforts to experiment with creative means to promote conservationist property uses and to manage environmental change in ways adapted to regional or local circumstances;
- expanding educational efforts to inform property owners not only about the purposes of new regulations but also about the easiest ways that they can comply;
- promoting mechanisms—including a free press, NGO participation in policymaking, and broader democratic discussion—to bring environmental problems to light in a timely way, so that transitional issues can be managed before push comes to shove.

Many of these suggestions entail political effort and compromise. But politics and compromise, even in the form of some limited or temporary accommodation with polluters, allow legitimate regulation to evolve while alleviating the immediate disappointed expectations that might otherwise become obstacles to progress. This is the nature of democratic change—and an important point for environmentalists to keep in mind as they push for the next generation of public health and ecological protection policies.

Property owners, for their part, should also keep an important point in mind: property claims are subject to change over time, and owners need to adjust their own expectations in light of common efforts to manage public resources. Owners are entitled to expect that new environmental regimes will carefully weigh costs and benefits, and they are entitled to expect fair *transitions* to new ways of managing environmental resources; but no one can expect that existing property uses will forever remain the same.

Property Concepts in Environmental Management

Beyond the management of property rights transitions, there are many exciting possibilities for using property concepts to further the protection of environmental resources. The 1990 Clean Air Act opened up an extremely valuable quasiproperty-rights experiment, with the effective privatization of a large portion of sulfur dioxide emissions in the United States and the creation of an emissions allowance trading program.[16] Similar efforts are now afoot to control water pollution throughout entire watersheds.[17] Such limited tradable emission rights have the virtues of traditional private property rights: they are bounded in scope; they allow a range of private choices; and they encourage thrift, planning for the future, and attentiveness to the rights of others. Anyone genuinely interested in environmental protection has to consider how such limited, legislatively created, propertylike rights might be deployed to preserve other environmental resources (see chapter 7).

Property rights—both private and public—are essential to a free enterprise system. The protection of both private and public rights is important because the goal of a free enterprise system, all other things being equal, is not simply to maximize the value of private goods. It is to maximize the value of *the sum of private and public resources.* Much of the recent public discussion of takings points out the dangers to private owners from uncompensated public expropriations. These dangers are real; public expropriations can unfairly single out particular private owners to pay for public benefits and, writ large, they mean that we could impoverish ourselves as a nation by discouraging enterprise and undermining commerce. That is why we have constitutional-ized judicial oversight of public regulation through the takings clause. As economists often point out, the threat of a compensation requirement makes legislatures appropriately cautious, and makes them think more carefully about the real costs and benefits of their proposals.

But when diffuse public resources come under pressure from private piggybacked uses, the failure to manage private owners' uses presents a differ-ent kind of legislative failure. That failure too can impoverish us as a nation, by decimating resources that are diffuse and difficult to turn into private property, but that are still immensely valuable to the public as a whole, now and (it is to be hoped) in the future.

Once again, it is important to recall that property and environmental pro-tection have the same goal: to increase the *sum* of well-being, both private and public. Both sides are a part of our "common-wealth," and property ownership and environmental law need to work together to further that common wealth.

Notes

1. Writings that are still central to our legal and political thought have long recog-nized this fact—John Locke's *Second Treatise of Government,* William Blackstone's *Com-mentaries on the Laws of England,* and James Madison's and Alexander Hamilton's *Feder-alist Papers.*

2. This is a point that is recognized even by such libertarian writers as Richard Epstein; see Richard Epstein, "Private and Common Property," *Property Rights* (1994): 17, 41.

3. See T. Anderson and P. J. Hill, "The Evolution of Property Rights: A Study of the American West," *Journal of Law and Economics* 18 (1975): 163.

4. The very term *public rights* historically reflected the fact that although a resource could not easily be privatized, it was nevertheless valuable to many people and subject to a kind of easement for public use. See Carol M. Rose, "The Comedy of the Commons: Cus-tom, Commerce, and Inherently Public Property," *University of Chicago Law Review* 53 (1986): 711; H. Scheiber, "Public Rights and the Rule of Law in American Legal His-tory," *California Law Review* 72 (1984): 271.

5. A careful look at nuisance cases, for example, reveals not a set of fixed substantive doctrines on what is and what is not a nuisance but rather a quite dynamic body of law, responding and permitting legislatures to respond when either private rights or the rights of the public came under threat.

6. *People v. Gold Run Ditch & Mining Co.,* 4 P. 1152, 1156, 1158–59 (Cal. 1884); *Rivers & Harbors Act of 1899,* ch. 425, sec. 14 (codified at 33 U.S.C. sec. 407 [1988]).

7. For example, in an effort to protect both private and public fishing rights, Massachusetts required nineteenth-century milldam owners to install rudimentary fish ladders in an unfortunately unsuccessful attempt to preserve Atlantic salmon runs (see Theodore Steinberg, *Nature Incorporated: Industrialization and the Waters of New England* [Cambridge and New York: Cambridge University Press, 1991]). Other legislation limited private owners' ability to create noise, smoke, and odors that inconvenienced the surrounding community.

8. For New York, see Louise A. Halper, "Nuisance, Courts and Markets in the New York Court of Appeals, 1850–1915," *Albany Law Review* 54 (1990): 301, 334–37; for the U.S. Supreme Court, see *Illinois Central R.R. v. Illinois,* 146 U.S. 387, 452–54 (1892).

9. For example, Chicago and Cincinnati passed smoke ordinances in 1881; see J. Laitos, "Legal Institutions and Pollution," *Natural Resources Journal* 15 (1975): 423; for the development of fish and game commissions in the later nineteenth century, see J. A. Tober, *Who Owns the Wildlife? The Political Economy of Conservation in Nineteenth-Century America* (Westport, Conn.: Greenwood, 1981), 179–254.

10. See, e.g., *Pearsall v. Post,* 20 Wend. 111, 135 (N.Y. Sup. Ct. 1838).

11. Information about environmental problems is itself a kind of "commons"—we ignore the effects of overfishing or of pouring wastes into rivers, for example, because we expect that if there is a problem, someone else will figure it out. If everyone thinks this, no one pays particular attention. See Carol M. Rose, "Environmental Lessons," *Loyola Los Angeles Law Review* 27 (1994): 1023, 1025, 1028.

12. In economic terms, the marginal costs of early uses may have been low—unlike the marginal costs of later entrants' added uses.

13. Unfortunately, many of the recent legislative proposals to relieve "takings" would make regulatory change more complex rather than easing transitions. Many add administrative hurdles to legislation, particularly to environmental legislation, or they attach a takings label to almost any drop in private land value, with the result that a vast range of regulation may be subject to costly and unproductive challenge. Such proposals would hamstring regulatory protections of public rights far more than was the case in traditional American law, and far more than is compatible with the normal character of property rights. See generally Zygmunt Plater, "Environmental Law as a Mirror of the Future," *Boston College Environmental Affairs Law Review* 23 (1996): 733.

14. For example, zoning ordinances typically exempt preexisting nonconforming uses, at least for some reasonable period of time.

15. See Ford Runge's call for agricultural subsidies to be converted to environmental performance payments in chap. 13.

16. The Clean Air Act itself, however, carefully provides that tradable emission rights are not technically "property rights."

17. Some water pollution reduction and trading schemes are described in William E. Taylor and Mark Gerath, "The Watershed Protection Approach: Is the Promise about to Be Realized?" *Natural Resources and Environment,* Fall 1996: 16.

four

Land Use

The Forgotten Agenda

John Turner and Jason Rylander

Take a look across America. From Boston to Baton Rouge, massive changes have taken place on the landscape and in our society. A seasoned traveler, dropped onto a commercial street anywhere in America, could scarcely tell the location from the immediate vista. A jungle of "big box" retailers, discount stores, fast-food joints, and gaudy signs separated by congested roadways offers no clues to location. Every place seems like no place in particular.

Hop in an airplane and look at the land use patterns below. Cul-de-sac subdivisions accessible only by car—separated from schools, churches, and shopping—spread out from decaying cities like strands of a giant spider web. Office parks and factories isolated by tremendous parking lots dot the countryside. Giant malls and business centers straddle the exit ramps of wide interstates where cars are lined up bumper to bumper. The line between city and country is blurred. Green spaces are fragmented. Only a remnant of natural spaces remains intact.

Powerful economic and demographic forces are at work in America. Population growth, migration, and fractured, low-density settlement and development patterns have altered the landscape. In little more than a generation this nation has been transformed—80 percent of everything built in the United States was constructed in the last fifty years.[1] Although much of this growth has been positive, the economic, environmental, and social costs of our current land consumption habits are now becoming increasingly apparent.

For much of America's history, expansion was a national goal. Immigrants were encouraged to settle the farthest reaches of the countryside.

Land was cheap and plentiful. In a nation so vast, the notion of resource scarcity took generations to gain credibility. But now the United States is a nation of 265 million people, with a population expected to increase by half again by the year 2050. Few places are unaffected by human development.

Increasingly, the nation finds itself struggling to meet the public's competing demands for open space, wildlife, recreation, environmental quality, economic development, jobs, transportation, and housing. Although it may never be possible in a democracy to meet each of these demands equitably, the tortured and fragmented way in which land use decisions are currently made all but ensures that conflict and crisis will continue to characterize environmental policy in the twenty-first century. It need not be so. A new land ethic must be developed, one that considers the needs of current and future generations, understands the carrying capacity of natural systems, and builds communities in which people can continue to prosper socially and economically.

Land use is the forgotten agenda of the environmental movement. In the past twenty-five years, the nation's many environmental laws addressed one problem at a time—air or water pollution, endangered species, waste disposal—and they have done it primarily through prohibitive policies that restrict private behavior. Although their achievements have been significant, such policies seem to offer diminishing returns.

Environmental progress in the next generation will increasingly depend on stemming the environmental costs of current land use patterns. Perhaps because *land use* is such a vague term, policymakers have difficulty grasping the linkages between the use of land and the economic, environmental, and social health of their communities. Environmental issues are traditionally debated in state and federal legislatures. Local governments and planning commissions consider land use. The next generation of environmental policymaking will require a more holistic approach—one that considers the impact of development on natural systems and integrates decisionmaking across political boundaries. It must build on the fundamental recognition that land use decisions and environmental progress are two sides of the same coin. So long as the cumulative effects of land use decisions are ignored, environmental policy will be only marginally successful in achieving its goals.

Past Patterns

For most of the last two centuries, Americans flocked to cities seeking a better life. But since 1950, people have begun to flee the urban core, moving out to fast-growing areas on the periphery. This outward migration has created a doughnutlike pattern of growth on the edges and decline in the center. Although the urbanization of America continues in the sense that more and more people are living within metropolitan areas or suburbs, the populations of many center cities have collapsed. Of the twenty-five largest cities in 1950, eighteen have lost population. Over the past forty years, central Baltimore and Philadelphia have each lost more than 20 percent of their residents, while Detroit declined roughly by half. St. Louis, the "Gateway to the American West," once boasted more than 850,000 people but now numbers only about 400,000 residents. During the same time, suburbs across the country have gained 75 million people, an increase of more than 100 percent. By 1990, more Americans lived in suburbs than in cities and rural areas combined.[2]

The suburbanization of America has consumed a tremendous amount of land. Metropolitan Cleveland's population declined by 8 percent between 1970 and 1990, yet its urban land area increased by a third. Even in cities that have not declined, their geographical reach has far outpaced population growth. The population of Los Angeles grew by 45 percent from 1970 to 1990, but the city's metropolitan area expanded 300 percent and now equals the size of Connecticut. Metropolitan Chicago grew in population by 4 percent, yet its developed land area expanded by 46 percent.[3]

Our land use patterns affect the environment in many ways. Most notably, development pressures have significant impacts on habitat. Even where forests and wetlands are preserved, new housing and commercial developments pave over open spaces, alter water courses and runoff flows, and rearrange scenic vistas. Our land use choices also impact air quality. For example, vehicle miles traveled by California's sprawling population have increased more than 200 percent in the past two decades as a consequence of distant suburbanization, exacerbating an already well-known smog problem in the region. Mass transit, which is only viable at relatively high population densities, becomes increasingly impractical as people spread out across the land.

Each year, another Paris—roughly 2.2 million people—is added to the American population. If current trends continue, 80 percent of these people will work and settle in edge cities and areas on the metropolitan fringe. Each new single-family detached home requires public services, schools, shopping areas, and roadways that further extend into farmland and open space. Coastal areas, the South, and the intermountain West face particularly acute growth challenges as more and more people, particularly retirees, migrate to these regions. Information technology makes remote locations more accessible, and a growing number of people who now can work from their homes are also moving for the natural beauty and personal security these places afford. Without comprehensive planning to address these demographic trends, patterns of explosive growth and voracious land consumption will continue with little or no consideration of the cumulative impacts on the environment and our future well-being. To ensure a reasonable standard of living for its people and a healthy environment, the United States must develop more rational and productive ways to manage resources—land as well as air, water, biological systems, and people.

Unfortunately, government policies have historically exacerbated trends toward separation and expansion. Land use planning in the United States has traditionally been the task of local officials who have used property zoning regulations and building codes as their principal tools. Zoning, a twentieth-century invention, was originally intended to protect property owners from their neighbors, to ward off economic, social, or environmental damage inflicted by adjacent land use. Although zoning has sometimes served these needs well, local planners have increasingly used zoning regulations to separate arbitrarily residential and commercial uses of land. As a result, the integration of shops and housing, narrow streets, and dense development that attracts admiring visitors to historic urban areas such as Georgetown in Washington, D.C., is prohibited by most local codes. Yet such multi-use urban development patterns offer residents more choices in type of housing, better access and convenience, less segregation by income and class, and a greater sense of community at far less infrastructure cost.

As a whole, the United States' land regulatory system is a failure. It is a policy of directed chaos—multiple programs and policies designed to address usually worthwhile goals but implemented in too small an area without regard to the health of the region and oblivious to their unintended consequences. As Aldo Leopold noted, "To build a better motor we tap the uttermost powers of the human brain; to build a better country-side we throw dice."

Land regulatory processes are often too narrowly focused, unevenly applied, and based on inadequate information. This promotes hostility among interest groups and leaves the general public with a sense of powerlessness and disenfranchisement. Most people are unaware or do not understand how land use decisions can dramatically affect their lives and neighborhoods.

Suburban jurisdictions often compete ferociously for business and development that once might have been located in the urban core. Municipalities lure businesses to their side of the border through tax breaks, infrastructure improvements, and other guarantees, but the costs of development, like increased congestion and pollution, are frequently borne by neighboring jurisdictions. With each county myopically focused on ways to increase its own tax base, the region as a whole becomes socially and economically fragmented. As jobs shift further from the central cities, people find they can live even further outside the metropolitan area and still have a reasonable commute to work. Those left behind in the older core cities, increasingly members of minority groups, face diminished job prospects, crumbling neighborhoods, and economic disparity.

The historical deference to local autonomy has, of necessity, precluded significant coordination among state and federal policies and actions. This disjointed approach has generated patchwork, ad hoc decisions. A basic challenge for land use policy in the future is to amend this approach to maximize environmental goals and to reflect a broader sense of community.

Transportation and housing policies have been major contributors to America's wasteful land use patterns. Transportation policies, designed almost exclusively for the automobile, greatly exacerbated suburban sprawl. Thousands of miles of trolley lines were abandoned or paved over to accommodate the car. The Interstate Highway Act of 1956 authorized construction of some forty-one thousand miles of new highways leading from cities to the hinterlands, and where the roads went, development followed. Business and suburban development flocked to the off-ramps of the new roads, but such growth came at the expense of cities and open space. The linkage between transportation and land use was rarely made, and national development patterns reflect that disconnect.

Federal housing policies also contributed to the growth of suburbia and the segregation of housing by class and race. In the decade following World War II, nearly half the houses built in the United States were financed with Federal Housing Administration (FHA) and Veterans Administration assistance. These programs boosted a construction industry floundering after the Great Depression and improved the U.S. stock of housing. But FHA-backed mortgages were only available for new homes—primarily single-family, detached houses on inexpensive suburban land. The agency did not support loans to repair, remodel, and upgrade older houses in the cities that might have provided affordable housing for growing minority and immigrant populations. Cities reaped few of the benefits of the postwar development boom.

Poorly designed statutes, including some of the nation's environmental laws, have had unintended consequences. The Superfund program, designed to promote the clean up of abandoned toxic waste sites, has failed to achieve its ends and may actually hinder the reuse of abused lands. Even in cases where costs would be lower to recondition an old facility, where infrastructure is already in place, lenders are reluctant to invest in such a project for fear of liability. The threat of liability for past contamination steers factories away from "brownfields" and encourages new development of "greenfields."

We are beginning to understand what we have lost. Despite tremendous technological advances, we have produced housing developments that demean rather than inspire our citizenry. We have built mile after mile of ugliness—cookie-cutter houses, subdivisions devoid of character, congested streets, commercial strips that assault the eye with garish signs and neon lights—all at the expense of townscapes, city cores, open space, productive farmland, and wildlife habitat. The costs of sprawl are not only aesthetic. The decline of cities and segregation of communities that result from land use decisions impose tremendous burdens on society. Local governments are increasingly aware that scattered large-lot zoning does little to protect habitat and often does not generate enough tax revenue to pay for municipal services. And the environmental costs of poor land use practices are rarely factored into local decisions.

Growth is inevitable, but ugliness and environmental degradation are not. With forethought, planners can channel growth to create more livable spaces and communities. Theodore Roosevelt called conservation a "great moral issue," and indeed our efforts to fashion a more sustainable society flow from a greater sense of reverence for the land and concern for present and future inhabitants. To pursue this ethic, we will need to identify more useful and understandable criteria for determining and measuring the costs of poor land uses. And we will need to overhaul conflicting government policies that inhibit sound land use decisions. To be successful, land use planning depends on good information and the support of people at all levels of government, the private sector, and the citizenry. The following principles offer an approach to guide thinking about land use issues for the next generation.

Think Systems

Better land use planning can only be achieved if citizens, resource professionals, and policymakers understand how development patterns impact natural systems. Long-term planning must consider systems—landscapes, watersheds, estuaries, and bioregions—to be sustainable. Analyzing and abiding by the carrying capacities of systems must provide the basis for the development of our communities in the future. Since natural systems often cross political boundaries, cooperative efforts involving federal, state, and local entities, including businesses and private landowners, are critically important.

Tomorrow's decisionmakers will need to draw from multiple disciplines and work with experts from many fields. Overspecialization yields myopic decisions—for example, roads built by traffic engineers who fail to consider the needs of pedestrians. Transportation planners, educators, recreational experts, financial experts, health providers, and government officials must learn to come together and trade valuable information in a public format with farmers, businessmen, water quality specialists, wildlife biologists, and environmentalists. A broader perspective is needed to assist communities in dealing with the diverse and complex issues affecting their lives.

Water quality and quantity, for example, are closely tied to the use of land and are of paramount importance to all people. Municipalities from New York to San Antonio are grappling with the need to protect open space and preserve water supplies in the face of increasing population pressures. But programs to protect and conserve water sources frequently extend far beyond city boundaries. In a case that illustrates the need for systems planning and regional cooperation, state and local officials in New York have jointly developed a plan to manage growth and development in the Catskill watershed to preserve the water source for New York City's nine million residents. With foresight and financial commitments, city, state, and federal officials are putting together a solution for the residents of New York that protects a larger land area, provides needed fresh water, and saves hundreds of millions of dollars that would otherwise have to be spent on water treatment facilities for the city. Systems thinking requires a thorough understanding of a watershed's limits and considers new development with that in mind.

Another example of the move toward a systems-based approach is the development of multispecies conservation plans to preserve threatened and endangered wildlife and plants. The Natural Communities Conservation Planning program in southern California is an experimental effort to preserve the state's remaining coastal sagescrub habitat in an area of high land values and growth demands. The complex and often controversial plan impacts five counties covering six thousand square miles and attempts to reconcile the conflicts between environment and development goals. Local, state, and federal partners are working cooperatively to carefully manage development, protect the California gnatcatcher and other imperiled species, and provide some long-term certainty for all stakeholders.

Engage Communities

Sound land use planning requires local knowledge, involvement, and spirit to provide the energy, staying power, and creative ideas that can come when neighbor joins with neighbor in trust to mold a collective vision for the future. Fundamentally, land use planning is community visioning. Without the input and involvement of local people, no plan can hope to succeed. This does not mean that locals can succeed on their own. State, regional, or federal involvement may be crucial in providing startup, technical assistance, guidelines, baseline information, and funding to help communities and multiple local jurisdictions plan for the future.

Although many people recoil from the thought of a federal land use policy, the reality is that the United States does have such a policy, albeit one that exists not by design but by default, arising from an uncoordinated collection of overlapping and often conflicting mandates and programs. Transportation policies, farm programs, disaster relief, water and sewer support, wetlands and endangered species laws, public housing, and financial lending programs combine to create a de facto national land use plan. An audit of federal programs affecting land use is long overdue to identify contradictions and move toward more consistent approaches to the use of land that complement regional and community goals.

Cooperation between governments is often difficult, but there are some models for integrating federal, state, and local needs. For decades transportation infrastructure programs at the federal level were developed without regard to local or regional land use objectives. The Intermodal Surface Transportation Efficiencies Act is a recent and innovative law that links transportation policy and investment with environmental concerns and local recreational needs, such as greenways and bike trails. Other models for cooperative land use planning at the federal level are the Coastal Zone Management Act and the Coastal Barrier Resources Act. A voluntary program, the Coastal Zone Management Act provides federal assistance to states that develop coastal management plans and ensures that subsequent federal actions will be consistent with the plans. The Coastal Barriers Resources Act avoids regulatory mandates but offers powerful disincentives by denying federal funds for roads, sewer plants, water systems, and flood insurance to developments that locate in sensitive coastal areas.

Fewer than a dozen states have comprehensive land use or growth management plans on the books, but those that do, such as Vermont and Oregon, have realized impressive results.[4] States can play a critical role in setting ground rules for local governments and assisting municipalities in grappling with land use issues like watershed protection that transcend jurisdictions. The ultimate objective of such plans is not to oppose growth but to ensure that development is consistent with community and regional objectives. Environmental policies can be explicitly built into these plans, rather than allowed to emerge incoherently as the function of thousands of disconnected land use decisions.

Perhaps the most significant achievements at the local level will not come from government but through the efforts of private citizens engaged in place-based conservation. Born out of frustration with national organizations or to promote a specific local issue, small grassroots conservation organizations have sprung up across the country. The proliferation of land trusts is enlivening the conservation movement with new energy and excitement. More than twelve hundred land trusts are now functioning across America—double the number of a decade ago—and their numbers increase almost weekly. These diverse and dynamic groups offer a fertile area for community ideas and involvement.

Use Better Information and Promote Education

In deciding what kind of land use strategy to employ, a community must understand its current makeup, strengths, limitations, and options. With today's information management technology, planners can review and interpret millions of bytes of data about soils, vegetation, water resources, biodiversity, view-sheds or vistas, tax structures, demographics, transportation and infrastructure needs, housing demands, recreation needs, and other local priorities. These systems are enabling community planners to develop models and make accurate predictions about the outcomes of policy choices.

In Florida, for example, the Conservation Fund in partnership with the MacArthur Foundation is using the technology of geographical information systems (GIS) in a facilitation process that allows planners and citizens in dozens of local jurisdictions to game possible growth management options for the region's future. In Alabama, in an unprecedented effort, seven major timber companies, working with Auburn University, are employing GIS mapping to gauge the effect of different timber practices on an entire watershed and test timber management strategies for their environmental impact.

Criteria must be developed for measuring the effects of land use decisions. Cost-benefit analysis can offer citizens and policymakers a better understanding of environmental and economic costs of land use. Quantifying the overall costs of sprawl would help communities assess how best to manage growth in their region.

For communities to take a lead in promoting sound land use policies, individual citizens will need a better understanding of the impact of land use choices on the environment and their future quality of life. Significant change will not soon occur in land use planning unless the public demands it. The more people understand these issues, the more likely a constituency will emerge for good land use planning. In short, we need to increase the ecological literacy of our citizenry. Ecological education at all levels should provide information about the relationship between the human environment and natural systems. Citizens must understand the inherent links of land use with clean air and water, safe and healthy neighborhoods, a prosperous economy, and a stable tax base if they are to be empowered to take action.[5] With information and education, communities can begin to develop the vision and leadership to build a more sustainable future.

Build Partnerships

Land use decisions are often controversial, but a growing number of enlightened leaders from various perspectives now recognize how much more can be accomplished when ideologies are checked at the door and rational people sit down to discuss solutions. Government, industry, nonprofit organizations, and citizens can have much greater impact working together than any one of them could have working alone. Next-generation policies must include new models of collaboration to avoid the rancor of our traditional adversarial approach to environmental issues.

Nowhere has there been more acrimony than in the debate over endangered species protection. Increasingly, however, private landowners, corporations, and the federal government are coming together to form habitat conservation agreements to protect imperiled species. These agreements provide certainty to landowners while ensuring an adequate level of protection for the affected species. In another example, the governor, environmental organizations, and timber companies in Maine sat down together to write a compact limiting clearcutting and improving forest practices across the state. The compact was placed on the ballot in the 1996 election as an alternative to a more extreme measure, and it passed by a wide margin. Such initiatives were unheard of just a decade ago.

In the next century, many significant gains in environmental quality will be the result of private sector initiatives. We must positively engage business leaders whose expertise, experience, political savvy, employees, and resources will be vital in addressing thorny issues such as nonpoint source pollution and biodiversity protection. Private landowners now hold most of the nation's remaining wetlands, endangered species habitat, timberlands, and open space. Partnerships between public officials, private groups, and major timber companies are already providing ways to harvest timber while expanding outdoor recreational facilities, reclaiming streams, and restoring habitat for threatened species.

We will also have to find ways to engage more private landowners in conservation. The U.S. Fish and Wildlife Service's "Partners for Wildlife" program, for example, provides funding and technical assistance to twenty-five thousand farmers and ranchers who want to restore wildlife habitat on their lands. Landowners get the satisfaction of improving the environmental quality of their property with assurance from the federal government that new regulations will not be imposed on their property should they choose to return the land to agriculture.

Empower the Disenfranchised

Resource professionals and environmentalists generally have failed to reach out to diverse social and economic groups in America. Often environmental quality is seen as an elitist issue of little concern to the poor. Many environmental activists have been slow to make the connection between declining cities and the loss of open space, between social issues and natural resource issues. Conservation should not be about preserving special places for the wealthy; it should be about improving the quality of life for all our citizens. We will never achieve success in conserving natural habitats if we ignore the human habitats crumbling in our midst. Poverty, joblessness, and unsafe streets are environmental issues.

The rise of the environmental justice movement has led to much broader participation. From clean water to lead paint to brownfield redevelopment, environmental concerns affect all Americans regardless of race or socio-economic position. The collaboration of people interested in environmental, social, and inner-city concerns will help change the way we think about land use issues. In the future, conservation, transportation, and development policies will thus take into account less affluent and less politically powerful members of society and galvanize inner-city groups to become active in environmental issues in their communities.

Nowhere is the cumulative effect of land use decisions more evident than in our cities. For example, because of a fear of crime and loitering, playgrounds, basketball courts, and community centers are often neglected or never built. "We are the first generation in history that fears its children," observes Charles Jordan, an African-American leader and director of Parks and Recreation in Portland, Oregon. "This can have a spiraling effect—less positive recreational opportunities, more anti-social behavior, more fear, and ratcheting down of the services we offer." Private citizens and church and civic leaders must begin to act on multiple fronts to counter the continued social stratification and decay of cities and urban people. It will take broad partnerships involving city, suburban, and rural interests. It will also take a broader understanding that the decline of cities, as well as rural areas, is everyone's problem. Everyone needs to be a part of the solution.

Protect and Enhance Wildness

Wildness is not a faraway place but a spirit—a characteristic of complex natural systems and places. Wildness might be found in a small woodlot, native grasslands, in a pond with tadpoles, or in a backyard visited by migratory songbirds. Wildness speaks of beauty, resilience, diversity, challenge, and freedom. It is one of the qualities that helps define us as a nation and uplifts us as a people. In protecting wildness, we are protecting something in ourselves. We sustain not just the tangible benefits of natural systems—new medicines, genetic materials for crops, air and water quality—but the character and durability of life itself.

Protecting wildness as a national policy was an American invention. We were the first country to establish national parks, forests, wildlife refuges, and scenic rivers. But there remains a need for more open and natural spaces in highly populated areas where little public land exists and outdoor recreation opportunities are few. A 1995 study conducted for a group of the nation's largest homebuilders found that Americans increasingly want to be able to interact with the outdoor environment in the places they live, through trails, woods, and open space. In that survey, 77 percent of respondents selected "natural open space" as the feature they would most like to see in a new home development.

Florida, Maryland, Missouri, Colorado, and other states have begun ambitious programs aimed at establishing greenways and protecting open corridors for wildlife and recreation. More needs to be done. Although voters repeatedly support bond issues to fund land acquisition, the lack of a coordinated constituency for public lands has allowed Congress to divert more and more acquisition funds away from the Land and Water Conservation Fund—an account established with revenues from offshore oil wells to meet federal, state, and local public land and outdoor recreational needs. This account should be restored to meet the nation's needs for the next century.

The rise of the land trust movement will reap tremendous benefits for open space while relieving public maintenance and acquisition burdens. These citizen-driven efforts offer one of the best hopes for conservation in the next century. We must experiment with new collaborative approaches like scenic easements, tax credits, estate tax relief, transferable development rights, and technical assistance to encourage the retention and restoration of as much wildness as possible.

Renew Spirituality

Conservation is sometimes difficult because it is in many respects a moral issue. It requires a sense of values, caring, and charity—a reverence for the blessings of nature and a shared commitment to the stewardship of the earth. We need not repeat the mistakes of this century in the next, but assuredly we will if we fail to take stock of our actions and accept responsibility for our land use choices. Our environmental affairs have become too secular. A renewed sense of morality and passion must be instilled in how we use the land and its products, how we care for one another, and what kinds of places we leave for future generations.

The best-selling author and philosopher Thomas Moore observes that the greatest malady of the twentieth century is the loss of soul. There is in America a growing disquiet, noted by many commentators, that the nation is losing its sense of purpose and morality. Loss of community, the "breakdown of society," is an oft-heard refrain. Our efforts to foster land stewardship and connect people to the land are an attempt to focus individual attention on common goals and values. Although people may differ in their religious, cultural, and ethical ideas, a sense of respect for nature is common to many traditions. It is time for a rediscovery of what we believe to be right and wrong.

America has always prized its individualism. In recent years, however, the individualistic strength that has driven countless pioneers in industry, communications, and finance has changed its character. Rather than strengthening community, our focus on the self comes at the expense of community. It seems to reject the notion that we have an ethical relationship to one another, to future generations, and to the land. Respect for one another is the precursor to ethical land use.

Changing the relationship of people to the land will not be easy. American laws governing land use have always been based on the premise that land is a commodity to be bought and sold for capital gain. Aldo Leopold put it most eloquently:

We abuse the land because we regard it as a commodity belonging to us. When we see land as a community to which we belong, we may begin to use it with love and respect. There is no other way for land to survive the impact of mechanized man, nor for us to reap from it the aesthetic harvest it is capable, under science, of contributing to culture. That land is a community is the basic concept of ecology, but that land is to be loved and respected is an extension of ethics. That land yields a cultural harvest is a fact long known but latterly often forgotten.[6]

Forging an ethical relationship with the land and its people is the challenge of our time. These principles offer only a guide for creating new tools and methods of decisionmaking that will shape the character of our national heritage. Improved land use policies will need to be based on a systems approach that reduces the waste of land and resources, enhances wildness and community character, permits growth and economic development, and preserves healthy and functioning ecosystems. "No net loss of greenspace" should be a goal for the twenty-first century. We must find ways to accommodate projected growth while preserving open space, farmland, watersheds, and rural communities. Redevelopment of brownfields and abandoned property must be afforded a higher priority than development of virgin lands on the metropolitan fringe.

We will need to develop more balanced, fair, and flexible regulatory approaches and reexamine government programs and procedures at all levels. Initiatives by local interests, public and private, must be encouraged, but additional leadership needs to come from state and federal governments, which can better coordinate actions that promote regional growth management objectives. Given that public support is the requisite for progress in the land use arena, we must make sure that local and statewide constituencies are developed, nurtured, and strengthened. Increased education and outreach to new partners are critical to success.

Land use planning is about people developing a sense of place and then deciding what their communities should look like in the future. It is not a radical idea, but it will require tremendous leadership, vision, and innovation. The payoff is a better quality of life, a stronger economy, and a healthier environment for America's future.

Notes

1. James Howard Kunstler, *The Geography of Nowhere* (New York: Touchstone, 1993).

2. Kenneth T. Jackson, "America's Rush to Suburbia," *New York Times,* 9 June 1996: E15.

3. Henry L. Diamond and Patrick F. Noonan, *Land Use in America* (Washington, D.C.: Island Press, 1996).

4. Other states, such as Florida, are experiencing explosive growth despite diligent efforts at state growth management planning. Florida grows by an average of 750 people each day, according to census figures, and each day on average the state sees 450 acres of forests leveled, 328 acres of farmland developed, and an additional 110,000 gallons of water consumed. Even under the growth plans developed by each county, Florida's population could soar from its current 14.5 million to nearly 90 million people by the end of the twenty-first century. See Leon Bouvier in "Florida's Growth Strains Services," *Washington Times,* 18 Nov. 1996: A10.

5. Many studies have detailed the high costs of suburban sprawl for municipal governments, which are hard pressed to pay for police and fire protection, schools, water systems, and sewers. A recent study done for Culpeper, Va., found that for every $1.00 in tax revenue from residential development the city must pay $1.25 to provide necessary services. The same study, conducted by the Piedmont Environmental Council, found that for every $1.00 in taxes collected from farms, forests, open spaces, or commercial lands, 19 cents was paid out for services. Large-lot-exclusionary zoning can be extremely costly, but many planners and citizens still cling to the notion that such practices are inherently profitable.

6. Aldo Leopold, foreword to *A Sand County Almanac* (New York: Oxford University Press, 1949), xix.

five

Sorting Out a Service-Based Economy

Bruce Guile and Jared Cohon

Factories spewing smoke into the air, sludge oozing into rivers, dumps strewn with debris—for most of us, these reflections of the dark side of industry symbolize the environmental impact of the modern world. Although such images can powerfully motivate public and private action, environmental policies and programs that focus on factories and extractive industries have become inadequate for addressing our nation's environmental problems.

Manufacturing, mining, and agriculture—three production activities often associated with environmental damage—account for less than a fourth of today's U.S. gross domestic product. Service businesses—from restaurants and stores to hospitals and airlines—account for more than three-fourths of the country's economy and 80 percent of its employment.[1] Any society's environmental problems are directly connected to its types and structures of economic activity—methods of production, patterns of consumption, and forms of economic organization. Changing patterns of production and consumption demand parallel changes in environmental policy. In particular, policymakers must recognize service businesses as core economic activities and thus as an important focus for the next generation of environmental analysis and programs.

Services are not necessarily more environmentally problematic than manufacturing, mining, or agricultural industries. But in an economy in which they dominate and serve as some of the most dynamic drivers of change, it is important to examine them directly in search of opportunities for both halting environmental degradation and improving environmental quality.[2]

Thinking about Services and the Environment

The service sector includes a diverse set of societal functions including:

- finance, insurance and real estate (18 percent of the 1992 GDP)
- wholesale and retail trade (17 percent)
- transportation, communications, and utilities (10 percent)
- health care (6 percent)
- business and legal services (5 percent)
- government services (12 percent)[3]

The environmental impacts of the service sector are as diverse as the components that make up this segment of the economy. The following four examples illustrate this diversity.

A handful of the largest discount general-merchandise chains in the United States, companies such as Wal-Mart, K-Mart, and Target, account for approximately one out of every ten dollars in national retail sales. These enormous retailers define, to a large extent, our environmental choices as consumers. Retailers make decisions about shelf space, location, presentation, and information regarding the environmental characteristics of products they sell. By their choices, they control our choices and set explicit and implicit environmental standards for products they choose to stock. They also have enormous power with manufacturers. The buying power of these category-dominant retailers makes their views on environmental impact of obvious concern to their suppliers. Their purchasing decisions effectively determine the success or failure of "environmentally responsible" products. Similarly, they have an enormous impact on the design of manufactured products as well as on the environmental characteristics of the manufacturing processes.

From plastic toys to sneakers and from dishwashers to automobiles, the final price paid by the consumer is far more than direct manufacturing costs. Your casual purchase of a compact disc illustrates the degree to which service businesses dominate manufacturing in economic and, therefore, environmental importance. Out of the $15.00 that you are charged, only about $1.20 pays for the labor and raw materials used to produce the CD. The rest goes for taxes (government services) and for developing, designing, transporting, marketing, and selling it. And each of these functions has an important environmental impact.

Omnipresent McDonald's illustrates the broad environmental impact of food service and delivery. McDonald's serves some twenty-two million meals each day, using two million pounds of potatoes as well as truckloads of meat, buns, and, annually, thirty-four hundred tons of sesame seeds.[4] The production process that McDonald's uses to source food, control quality, deliver supplies to restaurants, and deal with waste is complex and demanding. The company adds considerable value to the agricultural products that are the raw materials of its production process. Its volume purchases give it the power to alter the way potatoes are grown and steer are slaughtered. Its standard hours of operation at thousands of restaurants affect the demand for electricity and gas.

Another service industry with substantial, easily recognizable environmental impact is commercial real estate development. In negotiations with local governments and sources of finance, real estate developers play a decisive role in determining local and regional patterns of energy use, waste disposal, and transportation choice. The location and characteristics of office buildings, warehouses, shopping malls, housing developments, mobile home parks, and hotels are largely in the hands of developers. In a very direct sense, real estate developers—often with little effective control by local governments—create the environmental footprint of urban and suburban living.

In sum, the environmental consequences of service activities in the United States are great, varied, often technologically sophisticated, and largely determined by a set of private, for-profit business decisionmakers. Additionally, the environmental footprint of services extends into manufacturing, agriculture, and other natural resource industries since all are connected by the value chain which begins with extraction and manufacturing and ends with reuse or disposal. The tremendous impact of service companies on the environment can be divided into three basic categories: upstream leverage, where the service company influences its suppliers and others up the value chain; downstream influence, where the service company influences its customers toward the end of the value chain; and environmentally responsible production, which requires us to consider how the "production" of services can be done more efficiently.

Upstream leverage. As the effective purchasing agents of millions of consumers, service companies exert tremendous leverage over their suppliers. In passive rather than active terms, such firms can create a "market" for environ-

mental improvement. As we have seen, retailers such as McDonald's are powerful forces.[5] Toys R Us, a service company, has higher sales revenue than the
two largest toy manufacturers, Hasbro and Mattel, put together.[6] Airlines and
transport services like United Parcel Service are the primary customers of
equipment suppliers like Boeing. As major service providers, these companies have the economic clout to control the way their suppliers design, package, and ship products, and sometimes even where they locate their factories
and warehouses. Thus, the environmental impact of service businesses is
enormous, whether or not the companies are aware of it.

Downstream influence. Well-managed service firms typically work
extremely closely with their customers, who depend on them for many types
of environmental information. In turn, service companies often have early
insights into consumer tastes, preferences, and regional buying habits. In
other words, service industries play a key role in both satisfying and creating
consumer preferences for goods and services, including their environmental
dimensions. Customers browsing in Wal-Mart's environmentally designed
store in Lawrence, Kansas, are "informed" both directly and indirectly about
the environmental impacts of products and merchandising.[7] Home Depot
has instituted extensive efforts to provide "green products" to customers.[8]
Viewers of Disney's movie *Pocahontas* come for entertainment but leave with
a message which is, in part, environmental. Any hotel that offers guests the
resource-saving option not to have sheets and towels changed every single
day of their stay enhances consumer education about environmental issues.
Service firms also educate their peers in other service sectors. Bankers often
draw real estate investors' attention to environmental liabilities associated
with a building or lot. In sum, direct customer contact, in combination with
scale, allows service companies to play an important role in consumer education about the environment.

Environmentally responsible production. Although we usually associate
the notion of production with manufacturing, services too must be "produced," often in a process that involves multiple steps. Many service companies have complex production operations that—inefficiently run—can use
large amounts of energy and generate huge waste streams. Fast (and slow)
food chains feed millions and a single transportation firm, UPS, reportedly
handles more than 5 percent of the value of the U.S. GNP every day. The

Kinko's chain of copy centers has an annual total energy bill of around twenty million dollars. The largest users of paper are all in the service sector. Service company policies and corporate approaches to environmentally responsible production of services have a large, direct impact on the environment.

Although we can generalize about the impact the service sector as a whole has on the environment, it is exceedingly difficult to summarize the environmental opportunities and best policy approaches for its diverse members. In part, the work needed to understand the relationships of these industries to environmental outcomes has not been done. In addition, the range and complexity of service businesses defy easy generalization about sound approaches. Further, the environmental analytical community is just beginning to apply systems approaches to characterizing the impact of services, and links between such analyses and policymaking do not yet exist.[9] Three examples—logistics and distribution services, financial services, and health services—illustrate the diversity and complexity of environmental considerations related to service industries.

Logistics and Distribution: Changing the Structure, and Environmental Impact, of Manufacturing, Wholesale, and Retail

Logistics and distribution (moving and storing materials, components, and final products outside a factory's walls) represent one of the most rapidly growing service sectors, and one with profound implications for the nature, volume, and spatial distribution of environmental impacts. Logistics and distribution have both fueled and been fueled by fundamental shifts in the way manufacturing is done. The days are numbered (if not already gone) for the widget factory with large, on-site inventories of raw materials and subassemblies and an adjoining warehouse of finished widgets ready for shipping. Today the widget factory, which might be in Indonesia rather than Buffalo, receives its inputs within hours of using them and stores very few widgets on site. Its raw materials are stored in the hull of a ship, on rails, or on a truck; its warehouse is UPS, Federal Express, or the U.S. Postal Service. In fact, Fred Smith, the founder of Federal Express, has written:

> Few people seem to have noticed the economic significance of express carriers in important emerging management practices in distribution

and logistics. The concept of just-in-time inventory management, in which the materials of production are scheduled to arrive when they are needed in the production process, is a simple idea with enormous implications. In an economy focused on keeping inventory lean, air cargo transportation equipment and systems become "flying warehouses"— accessible, secure, 500-mile-an-hour storage facilities for those items somebody wants to use tomorrow.[10]

During one particularly busy night, for example, Federal Express moved 1.6 million packages through its Memphis hub. The energy cost of Federal Express's 450-plus aircraft and 31,000 motor vehicles is a significant part of the company's operating expenses and a key determinant of its environmental impact. From a life-cycle perspective, the business and technology decisions made by freight carriers represent a critical dimension of the materials and energy flows associated with this new structure of production.

In the mid- and late 1980s, the focus of most U.S. companies competing in global markets was on product development and manufacturing processes. Over the last decade, attention to what happened inside a factory's walls has spilled over into new approaches to vendor and materials management and into new approaches to forecasting, assessing, and responding to changes in demand ("reverse" logistics, demand-chain management, quick-response manufacturing). Rapidly evolving logistics and distribution systems are at the center of all these new approaches. In some cases, the development of new and successful distribution and logistics systems has been led by the manufacturers, but in other areas a new generation of powerful retailers or transportation service providers have moved "upstream" toward manufacturing and materials/goods purchasing.

During the late 1990s and early 2000s, a widespread network of factories connected by distribution and logistics operations will drive fundamental changes in the structure of production and consumption in the U.S. and global economies. Those changes will have profound, as yet only dimly perceived, implications for the environment. In particular, the technologies of transportation, storage, material handling, and information will link consumers and widely dispersed production centers through distribution systems that will be able to deliver goods and services almost anywhere very rapidly. As this happens, the value-adding chain for products and services will change in ways that dramatically affect the environment as well as the structure of production and consumption.

One clear qualitative impact of this logistics-based economy is a shift in the location of environmental impacts. From the perspective of U.S. policy-makers, point sources of pollution are traded for mobile sources when goods are made overseas and shipped here. From a broader spatial perspective—assuming no change in manufacturing efficiency or environmental impact—there may be a net increase in environmental impact: the mobile sources added to the same-point sources. However, more pollution from an increase in transportation at the manufacturing end might be offset by savings in the downstream product-distribution end.

The increased emphasis on logistics can result in potentially higher, and hidden, environmental costs of packaging. For example, one of the authors recently ordered sixty dollars' worth of supplies from a major office products company: a telephone, two paper tablets, an electric pencil sharpener, and several kinds of files. The telephone order arrived two days later in four large boxes. In one large box was the telephone in its own General Electric–branded box. The two tablets were in a separate box deep enough to contain twenty-five. Two additional large boxes held the balance of the order. Stacked, the boxes reached a height of about four feet. Unpacked, all of the order fit easily into the smallest box.

This incident raises several questions and observations. What warehouse management system, or network of warehouses, does the company use, in combination with an express courier, to deliver an order in two days at a very small price above what it would have cost to get the same items in a store? Most important, why was the system apparently so casual in its use of cardboard and in the volume of the shipment?

It is not clear what environmental policy prescriptions best address the types of issues raised by the changes in logistics and distribution. At the most abstract level, the answer is easy: policies should establish incentives or rules to reduce the long-term environmental impact across the system as a whole. At a more concrete level, unanswered and, possibly, unanswerable questions abound. How much and what type of regulation would be most effective in reducing the environmental impact of logistics and distribution activities? Should incentives favor concentration or dispersion of manufacturing? Of wholesale activity? Of retail? Even if we had the analytical ability to make a judgment on such matters, what incentives would work? In many cases, public investment in roads, airports, or other facilities determines the shape of transportation systems and provides a way of changing the local environmental impact. However, minimizing the local effect will almost certainly not minimize the broader environmental impact of the entire transportation system.

In the end, environmental policymakers are probably going to be forced into a goal-setting role. They might be able to evaluate the environmental impacts of the evolving system, but there will be little they can do to control it with traditional regulatory mandates. The wisest course may be to make wasting materials and energy, and other identifiable environmental harms the system creates, appropriately expensive. Then leave it to profit-seeking industry to shape a system that minimizes waste, energy consumption, and other impacts. Regulators will need to be alert, however, to the tendency of such incentives to drive participants to shift waste problems, emissions, or energy consumption to their suppliers or customers, to other industries, to other locations, or to society at large. A systems approach to environmental policymaking, perhaps using the framework of industrial ecology (see chapter 1), is one way to guard against such problems.

Financial Services: The Purest Form of Downstream Influence

In market economies, financial markets are the key means by which risks are communicated between actors. A hotel in Phoenix fails and the bank holding the developer's note refuses to lend to any more commercial real estate in Phoenix, no matter how good its prospects. A company announces less-than-expected quarterly earnings and its stock price plummets as the market reflects a new collective assessment of the risk/return ratio for its stockholders. In other words, financial markets rapidly adjust for the changing fortunes of an enterprise and serve as a mechanism for quickly communicating shared assessments of the probability of future economic scenarios.

To the extent that environmental issues pose financial risks or opportunities to corporations, financial markets would be expected to reflect those considerations. A company that risks a significant environmental liability as a result of its irresponsible manufacturing practices should pay more for its capital than a comparable company with more responsible practices. Local governments with environmentally problematic waste disposal plans and practices should have to issue bonds with higher interest rates than governments with environmentally sound practices. Better market performance by an environmentally progressive company should also be reflected in investor decisions.

However, the number of instances in which environmental issues are regarded as important enough to influence the availability or price of funds is quite small. Environmental liabilities attached to property and equipment have become visible enough to be explicitly considered by real estate lenders and equity investors. In corporate acquisitions the due diligence process always includes an assessment of the potential environmental liabilities being acquired along with the company and its assets. By and large, however, environmental costs and opportunities are inadequately reflected in capital markets.

Although it is unclear why financial markets fail to appropriately value environmental costs and opportunities, it appears to have something to do with the character and value of information and the way it is used in decision-making. The relationship between environmentally responsible practices and better financial performance is often not obvious to either corporations or their financial service providers. At the core of this problem is the fact that most companies—both in internal discussions and in the way they represent the company to outsiders—fail to treat environmental performance as a critical element of financial performance.

Most companies (much less their financial advisors) do not understand—on either a conceptual or empirical basis—the financial implications of many investments with important environmental consequences such as eco-efficiency investments.[11] The separation of environmental and financial reporting or the complete lack of reporting on environmental issues obviously influences a company's internal data collection efforts. It also means that there is little common ground for environmental and financial analysts. The language of finance (price-earnings ratios, cash flow, generally accepted accounting principles, and tax burden) is very different from the language of environment (emissions, chemicals, habitat). Finally, changing a company's environmental footprint can be a slow, difficult, and hard-to-measure process. Financial markets are poorly equipped to understand or value operational improvements that are slow to mature, hard to manage, and hard to measure. This is especially true when there is a conflict between long-term environmental and financial sustainability and short-term decisionmaking driven by concern for quarterly earnings.

Information is an obvious key to making financial markets work toward improved environmental outcomes. Yet, even if the measurement and interpretation issues cited above were solved, there would still be some serious complications. For example, who owns the environmental information? Could, say, the manager of a mutual fund disclose to potential investors environmental information from a company whose stock is in the fund? This complication and others derive from the complexity of today's instruments and markets, a consideration that must be reflected in policy.

A key question is whether the failure to recognize environmental considerations is a failure in the performance of financial markets and, if so, whether it justifies and is amenable to government action. With little congruence between financial and environmental measures of corporate performance and prospects, what are the likely environmental benefits of developing and forcing the use of more holistic measures? If government regulation forced every company to prepare and present a paired environmental and financial audit, what would be the realized, long-term gains in terms of environmental quality? And at what cost?

Blunt instruments such as the burdensome costs imposed by Superfund regulations have demonstrated that financial markets can be educated to consider environmental risks. Consumer choice of "environmentally friendly" products can align manufacturers' economic and environmental interests. Expanded use of eco-labels, perhaps even the development of a standardized "eco-facts" form, would be helpful. However, policies that "mainstream" the positive and negative financial impacts of environmental actions will probably be the most useful. Changing accounting standards established by the Financial Accounting Standards Board (FASB) to require faster and fuller acknowledgment of corporate environmental liabilities would contribute to the success of such policies.

Health Services: Unusual Challenges in Environmentally Responsible Production of Services

With regard to most production and consumption activities in an economy, a fairly simple logic applies to thinking about environmental impact. In production, incremental changes that maintain the value perceived by the consumer and reduce the environmental impact of a good or service are generally positive changes. In consumption, changes that reduce the material and energy "lost" to the environment during consumption—while maintaining the value perceived by the consumer—are generally positive changes. This simple logic is, however, confused by the special character of health services, by the sometimes inscrutable ways that costs are allocated, by the intimate relationship between medical services and public health, and by the different ways in which individuals and society as a whole address human suffering and life.

In no service sector are different interests more entangled. Health services are delivered by a complex and rapidly changing industry of large and small, competing and cooperative producers responding to demand that depends on the behavior of consumers in concert with insurance companies, public health officials, and education systems. Our health care system must deliver medical services, provide support for those who cannot pay, relieve suffering, and keep costs down—as well as minimize environmental impacts. Nevertheless, there are real opportunities to reduce the environmental burden of health services. In many cases, the current pressure to lower or contain health care costs is entirely consistent with reducing waste and energy use.

At the same time, changes in the industry driven in part by cost-containment efforts and in part by advances in information technology are creating new structures of production and consumption and new opportunities and challenges for environmental policymakers. For example: New electronic equipment now permits patients to be tested and monitored at home rather than requiring them to commute to hospitals or clinics. Similarly, digitized medical records and electronic imaging systems are being used to transmit information from one medical facility or professional to another, reducing the need for expensive, on-site storage space and the consumption of paper and other raw materials.

To cut costs and still provide better patient care, the health services industry is increasingly emphasizing the formation of integrated delivery networks or systems of services. As a result, decisions about how, where, and with what equipment health services are provided are being centralized and, to a certain extent, rationalized. The impetus is to achieve lower costs, but the result of this systems approach is a structure that is likely to be more amenable to efforts to lower the environmental impact of delivering health services.[12]

In the context of these changes, what can be done to make sure that environmental costs and benefits are adequately and sensibly considered in the health care sector? Some issues are common with other service businesses, such as hotels, warehouses, offices, and restaurants.[13] Other concerns reflect basic issues such as recycling versus disposal: is it better to autoclave and reuse than it is to employ only disposable instruments? Some, however, are unique to the industry, such as how to handle biomedical waste that sometimes is radioactive and often carries a risk of infection.[14]

Ultimately, there are many more questions than answers. What is the environmental impact of an infectious disease? How could (or should) environmental analysis play a role in evaluating trade-offs between investments in preventive care and acute care? What are the most important environmental impacts from the delivery of different types of health services? Should environmental cost—almost impossible to measure and difficult to estimate—play a significant role in cost-based rationing of medical services?

In sum, environmental concerns alone will not drive the delivery and consumption of human health services, nor should they. But the industry's push for cost containment and its rapid absorption of technology are opening new windows of opportunity. Medicaid, Medicare, large employers, insurance companies, pioneer facilities, and medical suppliers are all institutional mechanisms that can help bring about environmental advances. Professional associations like the Association of Operating Room Nurses—some forty-four-thousand-strong—can be forces for education and environmental progress.

The Challenge to Environmental Policy: Understanding the Environmental Impacts of Services

Two main lessons emerge from this chapter. First, if environmental policymakers focus solely on manufacturing, mining, and agriculture, there will be huge gaps in the solutions they design. Services dominate the U.S. economy. They are not just connective economic functions or tertiary industries. Service companies have been, and continue to be, important sources of environmental problems and of opportunities for environmental solutions. Their upstream leverage on suppliers and downstream influence on customers make service industries a potential fulcrum for next-generation environmental policy.

Second, the diversity and complexity of service industries must be reflected in an equally diverse and complex set of environmental policies. There is no single environmental policy tool to address services-related harms or opportunities. Particularly when factories are replaced by complex chains of production sites, transportation equipment, and cyberspace, incentives and information are likely to be more effective as a regulatory strategy than would be a command-and-control regime.

Environmental information is key. New environmental accounting methodologies should be incorporated into financial analysis and corporate decision-making. We also need better environmental indicators and data collection systems. But getting this done will require many issues and questions to be addressed. The question of who owns environmental information and how it gets used in financial markets is central.

Policymaking—in all its complexity and messiness—is a design problem that must address two basic questions. First, what needs to be done? Second, how can we then create an incentive, a regulation, or an organization that stands the best chance of having a positive impact? Further, policymaking is a remarkably robust enterprise even in the face of little agreement and poor information. In most cases, policymakers are uncertain of, or don't agree on, the exact nature of the "problem" being addressed. They are even less certain about the likely impact of policies or programs. Nonetheless, policy experimentation and genuine progress do occur. In policies to address the detrimental environmental impacts of service industries, however, the huge gaps in our knowledge may overwhelm the policy process.

The analytical community that supports policymaking needs to address and find answers to two sets of questions:

First, what is the environmental impact of a service industry? We really know very little about the environmental impact of, for example, industries such as telecommunications, food wholesaling, or outpatient care centers. What is the energy-use and materials-flow footprint of each industry? What are the boundaries for comparing one type of service delivery with another— for example, shopping in a regional mall versus buying by mail? What is a reasonable way to perform a life-cycle analysis of a financial product, of a rapid package delivery service, of an information storage system? Are such issues worthy of intense policy attention or are the environmental impacts of such industries a secondary concern?

Second, what is the relationship between the environmental footprint of a service business and the constraints and opportunities in business decisionmaking? How do decisions by businesses in service industries affect the materials flow and energy use of the manufacturing, agriculture, and mining industries? How valid is the notion that service companies have a "choice" to exercise upstream leverage or downstream influence rather than simply doing what is economically expedient? Is environmentally responsible production of services reliably and genuinely a sound economic decision or only a transient marketing ploy?

Part of the reason we have so little information on the environmental aspects of the service sector is that government oversight has concentrated on emissions, not resource flows, and on final disposition of wastes (especially toxic wastes), not transfers. This narrow focus has been mimicked by the environmental, engineering, and management communities, who have overestimated the analytical complexities of the service industries and underestimated their impacts. The overall result is that major direct and indirect environmental forcing functions are left unstudied.

It won't be easy to address these issues. But answers are necessary in order for us to move forward. Government can help strengthen incentives for developing environmental tools and a common language by establishing information standards, collecting and distributing useful data, and, in some cases, forcing the collection of information by companies. Policies that mainstream the environmentally relevant data, analysis, and experience are likely to be powerful agents of change, guiding the way that service companies look at environmental opportunities and risks.

Notes

1. In addition to reflecting discussions at a workshop at the Yale School of Forestry and Environmental Studies in February 1995, this paper draws heavily on the work of James Brian Quinn of the Amos Tuck School of Business at Dartmouth and on various projects of the National Academy of Engineering (NAE). In particular, B. Guile and J. B. Quinn, eds., *Technology in Services: Policies for Growth, Trade, and Employment* (Washington, D.C.: National Academy Press, 1988); B. Guile and J. B. Quinn, eds., *Managing Innovation: Cases from the Services Industries* (Washington, D.C.: National Academy Press, 1988); and J. B. Quinn, *Intelligent Enterprise* (New York: Free Press, 1992). Also, the participants in two NAE workshops on service companies and the environment—October 1994 in Washington, D.C. and July 1995 in Woods Hole, Mass.—played a crucial role in developing the arguments in this paper. Errors and omissions are the responsibility of the authors, but the credit for pioneering in this area of investigation must be shared widely.

2. The dominant role of services in the U.S. economy is not a new phenomenon—the

number of people employed in service businesses has exceeded the number of people employed in manufacturing and agriculture combined since about 1950, and service sector employment doubled between 1970 and 1995. However, misperceptions abound. Services are often regarded as secondary industries that are technologically backward, employ people only at low wages, and are not capital-intensive. In reality, the technological intensity of service industries is often high (transportation, telecommunications, and health care services, for example); many service industry incomes are often well above average (doctors, lawyers, investment bankers, and airline pilots, for example); and substantial amounts of capital can be required (transportation firms, communications firms, and national retail chains are excellent examples).

3. U.S. Department of Commerce, Economics and Statistics Administration, Bureau of the Census, *Statistical Abstract of the United States 1995,* 115th ed. (Washington, D.C., 1995), 452 and 779.

4. Quinn, *Intelligent Enterprise.*

5. The story of McDonald's conversion from polystyrene to paper-based quilt-wrap containers is described in S. Svoboda and S. Hart, *McDonald's Environmental Strategy,* National Pollution Prevention Center Document 93–3 (Ann Arbor: University of Michigan, 1993).

6. See Meeting Summary of the NAE Woods Hole Workshop on Technology, Services, and the Environment, Woods Hole, Mass., 1995.

7. See also D. M. DeKeysor and D. A. Eijadi, "Development of the Anderson Lighthouse for the Wal-Mart Environmental Demonstration Store," *Proceedings of the Second International Building Conference,* Special Publication 888 (Gaithersburg, Md.: National Institute of Standards and Technology, 1995), 143–51.

8. See D. J. Lober and M. D. Eisen, "The Greening of Retailing," *Journal of Forestry* 93, no. 4 (1995): 38–41.

9. Industrial ecologist Thomas Graedel of Yale University has begun to characterize service businesses with regard to their life-cycle stages. Personal communication, February 1997.

10. Frederick W. Smith, "Air Cargo Transportation in the Next Economy," in Guile and Quinn, *Technology in Services.*

11. See Stephan Schmidheiny with the Business Council for Sustainable Development, *Changing Course: A Global Business Perspective on Development and the Environment* (Cambridge: MIT Press, 1992).

12. The authors would like to thank Stephen M. Merz of Yale–New Haven Hospital for these specific examples.

13. This colorful description was offered by Claude Rounds of the Albany Medical Center.

14. A prime example is the Healthcare Resource Recovery Coalition (HRCC). The authors are grateful to Kathy Wagner for information about this industry collaborative effort.

Globalization, Trade, and Interdependence

Elizabeth Dowdeswell and Steve Charnovitz

Observations about globalization have become clichés. Yet the growing degree of international interdependence—both ecological and economic—has important consequences for the next generation of environmental policymaking, particularly as it affects U.S. domestic policy and as
the United States considers its role in a changing world. In recent years,
governments have increasingly chosen to join voluntarily in a world of free
trade, economic cooperation, and relatively open borders. They have not,
however, "chosen" an "open" environment. It is simply a fact of life on this
planet. Nations are environmentally interdependent because pollution
does not stop at national borders. Ozone layer depletion, climate change,
and radiation from nuclear accidents present inherently global risks.

In this chapter, we explore the challenge of managing the interdependence entailed in the cohabitation of our planet by almost 190 nations.
Understanding the connections among countries and the linkages between
environmental challenges is critical to achieving sustainable development.
Using the recent debate on "trade and environment" as a starting point, we
suggest that paying more attention to the implications of globalization can
improve national environmental programs and permit a clearer understanding of what it means to enjoy sovereignty in an interdependent world.
We also examine how the need to coordinate environmental efforts engen-
ders new institutional imperatives for global policymaking.

The Implications of Interdependence

Interdependence influences the problems nations face as well as their policy toolkits. The problems we face are shaped by the contradictory ten-
dencies of interdependence—to strengthen national security and welfare

in some ways while expanding vulnerability in other ways. Moreover, interdependence can frustrate unilateral action on the one hand and foster new types of intergovernmental cooperation on the other. Trade broadens the international dimension of the environmental challenge, creating competitiveness tensions that can work against sound environmental policies. Yet, international market forces can also have a positive effect by transmitting incentives for environmentally sound behavior.

The interaction of national economic and environmental policies is undergirded by an even deeper relationship: the connection between ecology and economy.[1] Although we have come to recognize that there are important links between these two spheres, we have been slower to perceive the reality that, for global issues, the spheres are ultimately one and the same. Perhaps an awareness of the common etymological root of the terms *ecology* and *economy* would keep us from forgetting this fundamental connection as we pursue the goal of sustainable development.

Trade and the Environment

Although it reflects just one facet of interdependence, the trade and environment debate provides a good window for observing how governments respond to globalization. Both trade liberalization and environmental policy can improve the quality of life and enhance social welfare. Yet such harmonious outcomes are not automatic. Economic policymakers must take account of ecological threats and potential health impacts. The failure to do so promises impoverishment, not enrichment. Witness the toxic legacy of communism in eastern Europe.

Economic and environmental policies can be mutually reinforcing. Environmental advances are easiest to achieve when a strong economy makes resources available for investment in pollution prevention and control. Environmental advances are hardest to obtain when poverty forces people into short-term decisionmaking. Poor people will cut down trees for firewood regardless of whether the loss of forest cover leads to soil erosion and other long-term environmental problems.

Some environmentalists oppose freer trade because they fear that economic growth will lead to increased production and consumption that creates pollution and increased pressure on natural resources. Correspondingly, some champions of development, convinced that the overriding imperative is the reduction of poverty, overlook or ignore environmental issues in their pursuit of expanded exports and what is conceived to be economic growth. We need policies that pursue simultaneously the benefits of sound environmental management and real economic development.

Since 1962, three major rounds of multilateral negotiations have led to freer trade. This opening and expansion of markets have provided, in general, large economic benefits to consumers. But it was not until the early 1990s that the issue of the environmental impacts of trade liberalization arose. Neither the postwar trade negotiations known as the General Agreement on Tariffs and Trade (GATT) nor its successor body, the World Trade Organization (WTO), has made significant progress in integrating environmental considerations into the trade domain. The most concrete advance is that there is now a greater appreciation of the links between trade and the environment. Following the 1986–1994 Uruguay Round of trade negotiations, for example, some countries conducted studies for the first time on the anticipated environmental impacts. Nations are beginning to realize that optimal trade policies cannot be set without taking environmental effects into consideration and vice versa.

The trade and environment debate has also highlighted the danger that environmental policy can be manipulated for protectionist purposes. Some requirements to include a specified amount of recycled content in consumer packaging or newsprint represent sound environmental policies. Other such rules pose trade barriers that are meant to disadvantage competing countries. Although there is no consensus as to how often environmental regulation is used as a guise for trade protectionism, there is widespread agreement that the environment is best served by policies that are not disguised efforts to advantage local producers. Just as enlightened trade policy has aimed to eliminate practices that enrich one country at the expense of another, enlightened environmental policy also needs to avoid mercantilist parochialism.

One key component of environmental policy centers on the problem of cost shifting—as economists say, the "externalization" of pollution or "free riding" by some on the environmental efforts of others. The spillover of air pollution, for example, from country A to country B shifts the cost of cleaning it up to country B (see chapter 7). The trade and environment debate has enriched environmental policy by injecting insights from how the trade

regime deals with the problem of cost shifting and free-riding by governments. In fact, such concerns became a central issue in the recent North
American Free Trade Agreement (NAFTA) debate; environmentalists feared
that lax environmental enforcement in Mexico would attract new industries
that generate greater transborder pollution. One of the strengths of the WTO,
and before it the GATT, has been a robust dispute settlement mechanism that
enables injured parties to complain about cost shifting and free riders. A central challenge for international policy makers is to develop similar structures
to control "beggar-thy-neighbor" environmental policies among countries.

International Investment and the Environment

A key lesson from studying trade and environment is that economic and environmental policies should be made in concert. This lesson also applies to the
link between environmental protection and investment flows. In the last fifteen
years, there has been an increased recognition of the impact of development
projects on the environment. The new dam that was once considered an industrial or energy matter is now recognized as an environmental matter as well.

As Stephan Schmidheiny and Bradford Gentry demonstrate in chapter 8,
private sector financial flows are a much greater factor in development than
public funds such as World Bank loans and foreign aid. This suggests a broader
point: that the responsibility for addressing globalization is not confined to the
international public sector. In addition to their effects on production and profits, financial flows also have environmental effects and represent important
opportunities to promote sustainable development. Governmental policy
frameworks are needed to help financiers, bankers, and insurers make investment decisions that are environmentally sound from a societal point of view.

The enhanced role of private sector financing also calls for new efforts to
figure out how to channel private international finance into environmental
infrastructure projects and to ensure that all privately financed factories,
roads, and other investments incorporate appropriate environmental protections. As foreign aid and multilateral development bank funding diminish in
importance, opportunities to inculcate environmental norms into economic
agreements and policies should not be overlooked. Likewise, the Multilateral
Agreement on Investment (MAI), a treaty being considered by Organization
for Economic Co-operation and Development countries to facilitate foreign

investment, could—and should—incorporate appropriate environmental safeguards.

Improving National Policies

In a global economy, there is heightened pressure for the adoption of efficient national policies. Because of international competition, the actions of government can become important factors in whether the companies in one country are more profitable than their competitors in other countries. Tax policies, for example, influence the rate of national saving and hence the availability of investment capital. Regulatory policies similarly influence the cost of production, and technology policies influence the rate of innovation.

In designing government policies, it is important to try to harness the forces of *competition* and *cooperation*. The benefits of cooperation are perhaps obvious. Efforts to address ozone layer depletion, diminished fisheries, and any number of other transboundary environmental problems require collaboration. But environmental progress can also build upon competition to achieve desired policy outcomes. By using a skillful combination of pollution taxes and emissions credits, for example, governments can use market forces to achieve a reduction in pollution at the lowest social cost.

Some environmental policy approaches blend competition and cooperation. The ISO 14000 standards prepared by the International Organization for Standardization (ISO) require companies that volunteer for certification to develop environmental management systems. Another voluntary program, the European Commission's Eco-Management and Audit Scheme (EMAS), has been running since 1995. Somewhat stricter than the more recent ISO 14000, it requires that companies establish and standardize environmental management and reporting systems and that they publish detailed public reports on their environmental management and performance. The objective of EMAS is to promote continuous environmental performance improvements. EMAS and ISO 14000 have tremendous long-term potential and will surely be emulated by others. Companies that complete either or both of these certification programs demonstrate in this way their good intentions toward the environment. Eco-labeling is another example of the fruitful interplay of competition and cooperation. Producers seek to qualify for an eco-label, highlighting their environmental virtues to customers, as a way of expanding their sales. The ISO standards, EMAS, and eco-labeling are becoming important benchmarks for measuring corporate environmental performance.

Cooperation is also important because nations face many of the same environmental problems. Governments can learn from each other about what policies work. Too often policymakers are unaware of the successful approaches used in other countries. The OECD was established in 1961 to help countries learn best practices from each other. Its useful work on chemicals and its recommendations regarding the polluter-pays principle and the role of economic instruments demonstrate the gains that are possible from symmetric approaches to problems.[2]

Although today's global economy offers unprecedented opportunity for trade and investment, the leap to participation can be daunting for developing countries. Opening a previously closed economy to competition forces change and creates some winners, but also some losers. Managing these tensions is an appropriate and neglected government role. Globalization policies need to be sensitive to how governments can smooth the transition process by helping workers and communities.[3]

Despite the increasing interdependence of the past twenty years, and the greater recognition of it, national policies often remain domestically oriented, disregarding impacts abroad and effects emanating from other countries. Another manifestation of parochialism is a lack of coordination among governmental ministries. For example, in many countries there is little daily interaction between the finance ministry and the environment ministry or between the trade ministry and the natural resource ministry. As a result, finance ministry officials, eager to achieve higher levels of gross domestic product and foreign exchange earnings, continue to advance their own too narrowly defined goals. They push timber companies, for example, to expand logging while giving little consideration to environmental consequences on forests and their indigenous inhabitants. Although the Earth Summit in Rio de Janeiro in 1992 got some of these ministries talking, further progress on integrating decisionmaking has been slow.

National Policymaking, Self-Interest, and Sovereignty

Some politicians and interest groups charge that stronger international rules will diminish national sovereignty. They often speak as if sovereignty were an unalloyed good. But no country can achieve its goals acting alone; international cooperation is required to maintain world peace, stem trade in narcotics, maintain exchange rates, control diseases, preserve the ozone layer, save whales, or do the hundreds of other tasks that the public expects of government. If sovereignty means merely that nations can choose policies without regard to others, then of course every nation can be sovereign (however feckless its actions). But if sovereignty means more—having the ability to accomplish the goals desired by the public—then governments must develop mechanisms to deal with the basic facts of interdependence.[4]

Mutual commitments in treaties can make participating countries better off. If a country resists international agreements that constrain national decisionmaking, then it will not be able to get other countries to abide by rules that protect their own stake in successful responses to global challenges. In brief, international agreements do not drain sovereignty. Instead, such agreements make it possible for countries to protect their own people. This point was well noted by the Permanent Court of International Justice, which declared in its first judgment: "The Court declines to see in the conclusion of any Treaty by which a State undertakes to perform or refrain from performing a particular act an abandonment of its sovereignty. . . . [T]he right of entering into international engagements is an attribute of State sovereignty."[5]

Of course, sovereignty does have a legitimate place as a policy goal. There should be no expectation of uniformity in economic and environmental policies among countries facing varying environmental circumstances and operating at different levels of development. But we should guard against the use of the term *sovereignty* as a shield for policies that serve certain vested interests at the expense of collective action in pursuit of a common good.

Looking ahead, we will probably see expanded efforts at international policy convergence through consultative mechanisms and treaty arrangements. Continuing developments in the European Union—such as the strengthening of environmental policy and the harmonization of standards— point the way to other programs of regional cooperation. Convergence will also occur through the private sector as businesses move to follow widely agreed upon environmental standards.

Establishing international rules and norms, however, is only one step. The agreements obtained must be implemented at the national level. Sometimes, even when states take leading roles in international negotiations, they may take a long time to ratify agreements, and an even longer time to translate international obligations into national legislation. Several of the countries that vigorously negotiated the Basel Convention on Hazardous Wastes and the Convention on Biological Diversity have yet to pave the way for national implementation.

In the 1970s, the U.S. government recognized similar procedural infirmities in the process for ratifying trade agreements. In response, trade officials devised the fast-track approval mechanism: federal legislation to implement a trade agreement gets an automatic Congressional vote without amendment after a strictly limited period of review. If the fast-track mechanism is available for multilateral trade agreements, perhaps it should be available for multilateral environmental agreements, too.

International Institutional Challenges

Many commentators are calling for a revamping of international environmental governance. They point to an insufficient response to fisheries depletion, chemical use, and climate change as evidence that stronger institutions are necessary. They also note the difficulty of coordination across environmental treaties. Another area of current focus is the size of UNEP. Some observers would like to see UNEP strengthened, but the recent trend has been reductions in the UNEP's budget.

Despite limitations, UNEP makes important global environmental contributions in a number of areas. It brings scientists together to make independent assessments of environmental problems at the global and regional levels. It catalyzes key environmental negotiations such as the Montreal Protocol on the Ozone Layer, the Basel Convention on the Control of Transboundary Movements of Hazardous Wastes and Their Disposal, the Framework Convention on Climate Change, and the Convention on Biological Biodiversity. It makes data available to environmental ministries, for example, through the Global Resource Information database. It furthers the development of international environmental law. UNEP has also developed model legislation for safe chemical use, and brokered the Global Plan of Action dealing with land-based sources of marine pollution.

A strengthened environmental regime—that is, the set of treaties, institutions, and practices operating together—would yield many benefits. Such a regime could help countries adopt more efficient environmental policies in the same way that the international trading system helps countries adopt more efficient trade policies. By enhancing its links to development policy, a stronger environmental regime could seek to prevent industrial and environmental policies from working at cross-purposes. Another need is to focus attention on the benefits industrial countries receive from the global commons and from environmental "services" provided by developing countries. The most important service may be forests that absorb greenhouse gases, thus mitigating climate change and serving as the habitat necessary to support biodiversity. Systematic accounting of these benefits might help lay the groundwork for integration of developing countries into the global economy.

Some observers have proposed the creation of a new Global Environmental Organization (GEO) to coordinate responses to global issues, facilitate exchanges on common problems, and reduce competitiveness tensions that result in suboptimal environmental policies.[6] But the prospect of anything like a GEO seems remote at this point. Therefore, strengthening the current structure may be the best approach available.

Several reform options have been proposed. First, building on its long-standing interactions with business groups and nongovernmental organizations (NGOs), UNEP could become a more effective counterweight to the commercial focus of the WTO. Second, mechanisms for long-term funding of international environmental investments shall be identified, including innovative forms of financing such as taxes on global pollution. Third, a new structure could be established to address environmental disputes.[7]

The WTO's Committee on Trade and Environment represents another potential environmental opportunity.[8] To date, the committee has not chosen to advocate new trade liberalization in support of environmental objectives. Key proposals by developing countries for greater attention to restraint on government subsidies for natural resource industries have made little headway. And the role of the trade regime in reinforcing utilization of environmental standards received virtually no attention within the committee.

The problem of coordination among international organizations goes far beyond the WTO. Although some specialized international organizations have proven effective, they often lack the capacity to deal with the connections between issues and the motivation to develop ongoing relationships with other international organizations. The need for coordination was recognized in Agenda 21 following the Rio Conference in 1992, but progress toward better linkages has emerged slowly.[9]

In particular, too little has been done to incorporate economic considerations and development concerns into international environmental policy-making. There are, however, some models to follow. The Montreal Protocol—the ozone layer protection program—phases out CFCs and other ozone-depleting chemicals over time to facilitate the shift to substitutes and to reduce the costs of compliance. It also sets different timetables for the phaseouts by industrial and developing countries and provides financial and technical assistance to developing countries seeking to fulfill their international environmental obligations.[10] These provisions make this treaty easier for all countries to accept, more likely to achieve its goals, and more durable.

We need to build on such approaches in negotiating future treaties. Innovation is called for in the areas of financial incentives, multi-tier obligations, phase-ins over time, technology transfer, use of economic instruments, and the assignment and marketability of property rights. More policy research into the design and evaluation of such mechanisms would assist policymakers in negotiating treaties that encourage wide membership and discourage free-riding behavior.

Another institutional strategy that should get more emphasis is the use of regional agreements. The European Commission, Association of South East Asian Nations (ASEAN), and Mercosur (a South American common market) are all improving their environmental programs. The environmental side accord of the North American Free Trade Agreement has engendered a new U.S.-Canada-Mexico Commission for Environmental Cooperation. Initial impressions are favorable as the commission performs its information-gathering, investigatory, and educational functions.[11]

The rapid expansion of regional economic integration provides a new opportunity for environmentalists. Just as regional trade agreements are used to harmonize trade and investment policies beyond what is politically feasible at the multilateral level, these regional agreements can also be used to test-drive programs of environmental collaboration and standards harmonization.

Finally, the role of NGOs in international development needs to be expanded. At the local level, NGOs can help transform the way individuals think about their own responsibility for sustainable consumption practices. At the international level, NGOs can bring information and competing ideas to governing bodies and international civil servants. UNEP and other institutions in the environmental regime have been among the most open to NGO participation of all international agencies. This openness can serve as models to other international agencies that are just beginning to grow out of their traditional government-only practices.

Although the pace of change in the global economy is very rapid, public institutions respond slowly. Despite increased recognition of environmental and economic interdependence, our modes of national and international governance have not evolved to address this new reality. The general public needs to be convinced of the national interest in *multilateral* solutions. Business leaders also need to recognize the new realities of environmental interdependence and to support efforts to address transboundary and global harms. Opinion leaders who promise to preserve national sovereignty by shunning international involvement must be challenged to deal with the facts of interdependence.

At the global level, international agencies cannot continue to maintain a reclusive approach that disregards the interconnectedness of issues, the interrelationships among agencies, and the increasingly expansive international civil society reflected in a diversity of NGOs. All international agencies must pay attention to the oceans, rivers, lakes, atmosphere, habitats, and biodiversity that sustain life on earth.

In conclusion, sustainable development requires a comprehensive perspective that integrates environmental, social, and economic goals. Governments should be prepared to introduce reforms that recognize the new economic realities and address past international policy failures. U.S. leadership is important in developing a consensus for carrying out these reforms. Governmental intervention should be set at a scale that matches the scope of the problem—be it local, national, regional, or global. Working with other countries, the United States should continue to look for better ways to manage the interdependence that is so important to its own and the world's prosperity.

Notes

1. See Donald Worster, *Nature's Economy: A History of Ecological Ideas* (Cambridge: Cambridge University Press, 1985), 36–38, 191–93 (discussing the etymology of *ecology* and *economy*).

2. Organization for Economic Cooperation and Development, *OECD and the Environment* (Paris: OECD, 1986); OECD, *Integrating Environment and Economy* (Paris: OECD, 1996).

3. See Dani Rodrik, *Has Globalization Gone Too Far?* (Washington, D.C.: Institute for International Economics, 1997).

4. See Abram Chayes and Antonia Handler-Chayes, *The New Sovereignty* (Cambridge: Harvard University Press, 1995).

5. The S.S. *Wimbledon* [1923] P.C.I.J., ser. A, no. 1, p. 25.

6. See Daniel C. Esty, *Greening the GATT* (Washington, D.C.: Institute for International Economics, 1994), esp. chap. 4.

7. See, for example, the *Report of the Foreign Policy Project,* a joint undertaking of the Overseas Development Council and the Harry Stimson Center (1997).

8. *The World Trade Organization: An Independent Assessment* (Winnipeg: International Institute for Sustainable Development, 1996).

9. United Nations Conference on Environment and Development, *Agenda 21* (Washington, D.C., 1992).

10. Duncan Brack, *International Trade and the Montreal Protocol* (London: Royal Institute for International Affairs, 1996).

11. North American Commission for Environmental Cooperation, *1995 Annual Report* (Montreal, 1996).

part two

Tools and Strategies for the Next Generation

seven

Market-Based Environmental Policies

Robert Stavins and Bradley Whitehead

It is not a new idea. Using market forces instead of bureaucratic fiat as a tool of environmental policy has been proposed by economists, discussed by policymakers, and implemented on a limited scale for two decades. But the concept of putting a price on pollution has yet to live up to its proponents' promises. Is this simply a breakdown between theory and practice? Has the effort to transform environmental regulations with economic incentives been nothing more than quixotic tilting at windmills? Should we continue to rely on more established—if costly—policy mechanisms? We believe the answer is no.

Market mechanisms can work. In fact, they have worked exceptionally well in a number of areas across the United States.[1] Of course, economic instruments, as they are sometimes called, are not panaceas. We have made less progress than we might have toward getting companies and individuals to pay for environmental harms they cause because of unrealistic expectations, lack of political will, design flaws, limitations in regulators' skills, and, all too often, obstacles thrown up by those who might be affected—in industry, the environmental community, and government. All of this can be addressed. Indeed, policymakers at all levels of government, in partnership with private businesses and nongovernmental organizations, should reinvigorate their efforts to develop and implement a next generation of economic incentives.

Properly designed and implemented, market-based instruments—regulations that encourage appropriate environmental behavior through price signals rather than through explicit instructions—provide incentives for businesses and individuals to act in ways that further not only their own financial goals but also environmental aims such as reducing waste, cleaning up the air, or reducing water pollution. In most cases, market mechanisms take overall goals of some sort—say, the total reduction of emissions of a specific pollutant—and leave the choice of how to accomplish this up to the individuals or companies concerned.[2]

In contrast, conventional approaches to regulating the environment, so-called command-and-control regulations,[3] typically force everyone to implement the same pollution control strategies, regardless of the relative costs to them of this burden.[4] For example, a regulation might limit the quantity of a pollutant that a company can release into the atmosphere in a given time period or even specify, in effect, that a certain type of pollution control device must be put in place. But holding everyone to the same target or mandating the same abatement equipment can be expensive and, in some circumstances, counter-productive. Thus, although this command approach has often succeeded in limiting emissions, it frequently does so in an unduly expensive way. Inevitably, it fails to tailor the demands imposed to the particular circumstances of each company. There is little or no financial incentive to do better than the law requires or to develop and experiment with new technology and equipment that might lead to even greater improvements in performance. The net result is a drag on productivity and complaints about regulatory inefficiency, both of which undermine commitments to achieving environmental gains.

Market-based instruments align the financial incentives of companies with environmental objectives. They can be cost-effective and can provide a powerful impetus for companies to innovate and to adopt cheaper and better pollution control technologies.[5] This leaves more room for economic growth or for more stringent environmental standards to be adopted.

Types of Market Mechanisms

Market-based instruments used in environmental programs can be divided into six major categories:

Pollution charge systems assess a fee or tax on the amount of pollution that a company or product generates.[6] Such "green fees" should be calibrated to actual emissions rather than simply to pollution-generating activities: for example, a charge per unit of sulfur dioxide released by an electric utility, not a charge per unit of electricity generated. Consequently, it is worthwhile for the utility to reduce pollution up to the point at which the cost of doing this equals what it otherwise would pay in pollution charges or taxes.

How it does this and how much it can reasonably spend until costs exceed the pollution tax will vary enormously among firms due to differences in their production designs, physical configurations, ages of assets, and other factors. The end result will be a substantial savings in the total cost of pollution control, as compared to forcing all firms to reduce pollution to exactly the same level or to employ the same equipment.

Setting the amount of the tax is, of course, not a trivial matter. Policymakers cannot know precisely how firms will respond to a given level of taxation, so it is difficult to know in advance precisely how much cleanup will result from any given charge. Nevertheless, in recent years, tax or green fee programs have been used successfully to phase out production of CFCs and other ozone-layer-harming chemicals and to promote better municipal solid waste management practices by charging people "by the bag" for the garbage they throw out.

Tradable permits get much the same results as pollution charges, but avoid the problem of trying to predict the results.[7] Under this system, policymakers first set a target of how much pollution will be allowed for an industry, an area, or a nation. Companies generating the pollution then receive (through free distribution or auction) permits allowing them a share of the total. Firms that keep their emission levels below the allotted levels can sell their surplus permits to other firms or use the allotment for one of their facilities to offset excess emissions in another one of their plants. Firms that run out of allowances must buy them from other companies or face legal penalties. In either case, it is in the financial interest of the participating firms to reduce emissions as much as they efficiently can.

There are now in place a number of successful applications of trading programs. In the 1980s, the EPA developed a lead credit program that allowed gasoline refiners greater flexibility in meeting emission standards at a time when the lead content of gasoline was being reduced to 10 percent of its previous level.[8] If refiners produced gasoline with a lower lead content than was required during any time period, they earned lead credits that could be either banked for the future or traded immediately with competitors. The EPA estimated that, compared to alternative programs, the lead banking and trading program saved the industry (and consumers) about $250 million per year and accelerated the phase-down of lead in gasoline.

A tradable permit system is the centerpiece of the acid rain provisions of the Clean Air Act Amendments of 1990. The law sets a goal of reducing emissions of sulfur dioxide (SO_2) and nitrogen oxides (NOx) by ten million tons and two million tons, respectively, from 1980 levels.[9] As discussed in more detail in chapter 14, electric utility companies annually receive tradable allowances that allow them to emit a specific amount of sulfur dioxide. Those that reduce their emissions below the level of their allowances can sell their excess permits. A robust market for the permits has emerged with savings estimated to be on the order of $1 billion annually compared to command-and-control regulatory alternatives.[10]

In another case, more than 350 companies in southern California are now participating in a tradable permit program intended to reduce nitrogen oxides and sulfur dioxide emissions in the Los Angeles area. The Regional Clean Air Incentives Market (RECLAIM) program operates through the issuance of tradable permits that specify and authorize decreasing levels of pollution over time. As of mid-1996, participants had traded more than 100,000 tons of NOx and SO_2 emissions with a permit value of more than $10 million.[11] Authorities are now considering expanding the program to allow trading between stationary sources (facilities) and mobile sources (cars and trucks).

Deposit-refund systems are familiar to many consumers because of the nine state "bottle bills" that have been implemented to reduce waste from beverage containers. Consumers pay a surcharge when purchasing potentially polluting products and get it back when the product is returned for recycling or proper disposal.[12] Although beverage-container deposits are the most common application, a few states have initiated deposit-refund systems for lead-acid batteries and other items.

Reducing market barriers can also help curb pollution. Measures that make it easier to exchange water rights, for example, promote more efficient allocation and use of scarce water supplies.[13] California, in particular, has achieved considerable improvements in water allocation by creating a market in water rights.

Eliminating government subsidies can promote more efficient and environmentally sound resource consumption and economic development. Below-cost timber sales, for instance, encourage overlogging. Similarly, federal water projects that provide below-market-cost water for farmers in California's Central Valley encourage wasteful irrigation practices and discourage water conservation. In these cases, market prices would deter waste and promote better environmental practices.

Finally, *providing public information* can improve environmental performance by allowing consumers to make more informed purchasing decisions and creating incentives for environmental care among companies. The Toxic Release Inventory, revealing emissions to air, water, and land of a large number of waste products, has emerged as a powerful tool for encouraging companies to reduce their emissions.[14] And the "dolphin-safe" label on cans of tuna fish virtually eliminated from the U.S. market tuna caught with methods that resulted in incidental, but significant, dolphin mortality.[15]

Barriers to Implementation

Notwithstanding considerable success in implementing specific programs, economic instruments represent only a small share of new regulation and a trivial portion of existing regulation. We must ask why market mechanisms seem to have achieved so little penetration. The most obvious reason is that there has not been a great deal of new environmental regulation. The Clean Air Act and Safe Drinking Water Act are the only major environmental regulations that have been reauthorized since 1990. And even when Congress has been willing to consider market-based instruments for creating *new* regulation, it has not been willing to substitute the technique for the *existing* regulations that now cover 14,310 pages of the Code of Federal Regulations. At the same time, most EPA employees were hired to oversee traditional command-and-control programs and some may be hesitant to switch courses. Traditional regulatory programs require regulators with a technical or legal-based skill-set. Market-based instruments require an economics orientation.

Many environmental organizations have also been hesitant to move regulation toward market-based instruments. Some groups worry that increased flexibility in environmental regulation will lower the overall level of environmental protection. Others believe that market mechanisms condone the "right to pollute" and that conventional government mandates thus have superior moral virtue. Finally, some environmental professionals, like their government counterparts, are simply resisting the dissipation of *their* experience and existing skills in dealing with command-and-control programs.

The ambivalence of government officials and environmentalists is mirrored by the regulated community. Many industries and companies have applauded market-based instruments in an abstract sense because of their promise of flexibility and cost-effectiveness.[16] As a practical matter, however, the vast majority of businesses have not enthusiastically lobbied for the implementation of these instruments. Much of the hesitation stems from reluctance to promote any regulation, no matter how flexible or cost-effective. Perhaps seasoned by experience, businesses fear that implementation might not prove as cost-effective as promised or that the ground rules could change after programs get under way.

From a political economy perspective,[17] private firms are likely to prefer command-and-control standards to (auctioned) permits or taxes because standards produce economic rents,[18] which can be sustainable if coupled with sufficiently more stringent requirements for new sources. In contrast, auctioned permits and taxes require firms to pay not only abatement costs to reduce pollution to a specified level but also costs of polluting up to that level. Command-and-control standards are also likely to be preferred by legislators for several reasons: the training and experience of legislators may make them more comfortable with a direct standards approach than with market-based approaches; the time needed to learn about market-based instruments may represent significant opportunity costs; standards tend to hide the costs of pollution control while emphasizing the benefits; and standards may offer greater opportunities for symbolic politics.

Moreover, those who would differentially be affected may be expected to press for changes. For instance, several high-sulfur-coal-producing states attempted to skew the acid rain trading program by forcing companies to install high-cost scrubbers instead of shifting to more economical low-sulfur coal from other states. At the same time several midwestern coal-burning utilities demanded—and received—"bonus" allowances. Additionally, for companies that have invested tens of millions of dollars in meeting existing pollution control requirements, any change in policy might entail more expense or the writing off of capital stock now in place. Indeed, for businesses to optimize their environmental investments, regulations have to be not only flexible but also predictable over time.

Coupling concerns about consistency with the antiregulatory climate pervading the country, many corporations have concluded that it is better to argue against *any* regulation rather than for better regulation. Several environmentally sensitive industries now argue in favor of voluntary industry programs rather than compulsory regulations. The chemical industry, for example, has developed Responsible Care codes that it says obviate the need for intensive regulation. The petroleum and paper industries have similar initiatives. The energy being directed toward these programs has diverted attention away from economic incentive approaches.

Part of the problem with market mechanisms is that the benefits they bring are often invisible to consumers, while the costs they impose as fees or taxes are all too plain. It is not obvious, for example, that gasoline and electricity prices are lower than they might otherwise have been because we successfully used market-based programs rather than command-and-control mandates to phase out lead and to reduce acid rain. On the other hand, long-distance drivers will pay more with higher gas taxes—and they know it. It is difficult to generate enthusiasm for economic instruments among those for whom it clearly means money out of their pockets.

Companies, moreover, often do not have internal incentive systems in place to reward managers who take advantage of market-based instruments. In many corporations, environmental costs are not fully measured and are not charged back to the business units from which they are derived. Moreover, the focus of many corporate environmental officers has been primarily on problem avoidance and risk management rather than on the creation of opportunities to benefit competitively from environmental decisions. Until corporate culture changes, the full potential of market mechanisms' cost-effectiveness and improved incentives for technological change will not be realized.

Next-generation market mechanisms. The limited use of economic incentives to date should not cause us to abandon or deemphasize market-based instruments as a next-generation policy option. Rather, we should make price signals a central part of the environmental policy toolkit. With more than $140 billion being spent annually in the U.S. on pollution control and cleanup, environmental policymakers need to seek more effective tools to maintain and improve environmental quality in a cost-conscious manner. This need dictates that we not lose the opportunity of using programs that can reduce costs and stimulate the development of new, more efficient technologies. In the long term, public support for environmental programs depends on confidence that the money invested is delivering good returns.

The first step toward better acceptance of market mechanisms is to improve the design of the programs. This must be done to counter the resistance of private firms, to calm fears of environmental groups and others about back-sliding on results, and to ensure that the actual cost savings come closer to, if not match, predictions. Accomplishing this means recognizing that market-based instruments are not a solution to all environmental problems. Rather, they are a useful element in what should be a portfolio of policy instruments. Indeed, some environmental problems will continue to require command-and-control solutions. On the other hand, market forces acting alone or voluntary industry initiatives may be sufficient to address other problems. But when regulation is called for, getting price signals to reflect environmental harms should be the first option considered.

An overarching design goal should be to make regulatory programs based on economic instruments more predictable. This requires stable rules, careful calibration of pollution control targets, and credible commitments to keeping programs in place for the long term. In addition, market-based instruments should be designed to deliver the greatest cost savings possible. Transaction and administrative costs must be reduced. Rights bestowed under these programs must be protected. Competitive market conditions must be maintained. The incentives for participation must be clear. When knowledge about environmental harms changes or new political pressures necessitate revisions to a market-based program, the transition should be made in a manner that does not detract from the program's efficiency.

In addition to design changes, the use of market-based instruments should evolve beyond Washington to the state and local levels. Although federal spending for environmental control continues to outpace state spending (in 1991, federal spending was about $18.2 billion for environmental and natural resource programs, compared to state spending of $9.6 billion), the gap is closing.

One of the most exciting uses of market-based incentives on the state and local level has been in an area not usually regarded as environmental: the general permitting process. A great challenge for state and local governments—and a source of frustration for new and growing companies—is the time required to issue permits for activities such as zoning, construction, and pollution discharge. Some states have developed programs that incorporate incentives into the existing framework for permits and inspections. For example, expedited evaluations of permit applications are often completed for firms that choose to participate in new pollution prevention programs. Although not a market-based instrument in the strict sense, such initiatives embody the spirit of what is called for in next-generation environmental policy: a relatively simple way to give firms incentives to meet environmental goals.

Market-based instruments can also be used to address the environmental issues at which most state and local initiatives are directed: waste management, land use, and air quality improvement. At the core of most municipal solid-waste problems, for example, are price signals that fail to convey to consumers and producers the true costs of waste collection and disposal. In fact, these costs are frequently embedded in property or other taxes. Some municipalities do highlight a charge for waste collection in their semiannual property tax assessments. However, since such charges are usually flat fees that do not vary with the quantity of waste generated by individuals, there is no incentive for users to reduce the waste they create. Unit pricing corrects this. By charging households for waste collection services in proportion to the amount of refuse they leave at the curbside, unit pricing ties household charges to the real costs of collection and disposal. Households thus have an incentive to reduce the amount of waste they generate either by changing what they buy, reusing products or containers, or composting yard and garden material. Moreover, if municipalities charge extra for unseparated refuse, they can also give residents an incentive to separate the recyclable components of their trash.

Unit charges will not solve all solid-waste management problems. They are difficult to apply to apartment units. Some form of "lifeline" pricing is required for low-income families so that these households do not pay a disproportionate amount of income for trash collection. And illegal dumping can be a problem if the programs are not organized properly.[19] However, this approach combines cost-effectiveness with minimum inconvenience. The number of these programs has mushroomed from one hundred in 1989 to some three thousand today.[20]

Market-based instruments can also help balance local economic growth with environmental protection of the land. As economic and population growth continues, a larger share of environmental problems will be associated with tensions over land use. Land-oriented tradable permit programs have already been adopted in several states, including New Jersey, Florida, and California. Florida established a wetlands-mitigation banking program in 1993 that allows the state and five local water management districts to license owners of wetlands property as "mitigation bankers."[21] Private developers are asked to offset the potential environmental damage arising from a proposed development by purchasing a "credit" from the bankers, who in turn agree to preserve and often improve their wetlands. Thus, those who diminish the amount of wetlands through development provide the resources to expand wetlands elsewhere in the ecosystem. Even before the program was formally established, a group of entrepreneurs set up Florida Wetlandsbank, which sells mitigation credits for forty-five thousand dollars per acre and uses part of the proceeds to improve degraded wetlands.

While working to incorporate the use of market-based instruments on the state and local level, policymakers should also work toward adopting new incentive programs on the federal level. In the hazardous waste area, deposit-refund programs could provide incentives not only to reduce the amount of waste but also to change disposal systems. The amount of lead, mostly from batteries, that enters landfills and incinerators may still be a significant hazard, despite EPA regulation of landfill construction and incinerator operation. The number of such batteries recycled each year has been declining. More than twenty million enter the waste stream annually, and this number could increase by some 30 percent by 2000. Under a deposit-refund system, a deposit would be collected by the administering agency at the time manufacturers sell batteries to distributors or manufacturers, who would pass on the charge to vehicle purchasers.[22] In time, the used batteries would be returned to redemption centers that would refund the deposit and then be compensated by the agency. Although some states have launched these programs, federal action is preferable when a national market or scale economies argue for a single system.

Market mechanisms may also be useful at the global level, especially in response to problems arising from diffuse sources. If, for example, the United States decides to participate in a binding international agreement to reduce worldwide greenhouse gas emissions, a carbon tax may be the most effective and least costly way to meet any emissions reduction targets. By altering price signals through charges based on the carbon content of fuels and tax credits for those establishing new carbon "sinks," a market-based regulatory system would internalize the potential costs of climate change. Higher prices would reduce demand for fossil fuels, thereby reducing emissions of carbon dioxide, and would stimulate the development of new technologies that are less carbon-intensive. Moreover, a properly designed revenue-neutral tax policy, under which carbon charges are offset by the reduction or elimination of payroll or other taxes, could help to protect the environment, reduce distortions associated with other taxes, promote economic growth, and render the program of greenhouse gas emissions controls more politically palatable.

By shifting organizational mindsets, developing new and needed skills, and overcoming the resistance of sometimes competing interest groups, we can make market-based instruments work for our collective benefit and bring environmental policy into the twenty-first century. If cost-effective regulation is a serious priority for environmental policymakers—and it must be in our world of tight budgets, both private and public—we cannot afford to overlook the opportunity to deliver more bang for the buck by harnessing market forces to protect the environment.

Notes

1. Janet Milne and Susan Hasson, *Environmental Taxes in New England: An Inventory of Environmental Tax and Fee Mechanisms Enacted by the New England States and New York* (South Royalton: Environmental Law Center at Vermont Law School, 1996).

2. See, for example, Robert Stavins, ed., *Project 88-Round II Incentives for Action: Designing Market-Based Environmental Strategies* (Washington, D.C.: Government Printing Office, May 1991); and Robert Stavins, ed., *Project 88: Harnessing Market Forces to Protect Our Environment* (Washington, D.C., December 1988). Both studies were sponsored by Sen. Timothy E. Wirth, Colorado, and Sen. John Heinz, Pennsylvania.

3. There is something of a continuum from a pure market-based instrument to a pure command-and-control instrument, with many hybrids falling between. Nevertheless, for ease of exposition, it is convenient to consider these two fundamental categories. See Robert Hahn and Robert Stavins, "Incentive-Based Environmental Regulation: A New Era from an Old Idea?" *Ecology Law Quarterly* 18 (1991): 1–42.

4. For a detailed case-by-case description of the use of command-and-control instruments, see P. R. Portney, ed., *Public Policies for Environmental Protection* (Washington, D.C.: Resources for the Future, 1990).

5. For an empirical analysis of the dynamic incentives for technological change under different policy instruments, see Adam B. Jaffe and Robert Stavins, "Dynamic Incentives of Environmental Regulations: The Effects of Alternative Policy Instruments on Technology Diffusion," *Journal of Environmental Economics and Management* 29 (1995): S43–S63. This paper develops a general approach for comparing the impact of policies on

technology diffusion and applies it to the most frequently considered policy instruments for global climate change.

6. A. C. Pigou is generally credited with developing the idea of a corrective tax to discourage activities that generate externalities, such as environmental pollution. See A. C. Pigou, *The Economics of Welfare,* 4th ed. (London: Macmillan, 1938). For a modern discussion of the concept and a number of case examples, see Robert Repetto et al., *Green Fees: How a Tax Shift Can Work for the Environment and the Economy* (Washington, D.C.: World Resources Institute, 1993).

7. See Robert Hahn and Roger Noll, "Designing a Market for Tradeable Permits," in *Reform of Environmental Regulation,* ed. W. Magat (Cambridge, Mass.: Ballinger, 1982). Much of the literature on tradable permits can actually be traced to Coase's treatment of negotiated solutions to externality problems. See generally Ronald Coase, "The Problem of Social Cost," *Journal of Law and Economics* 3 (1960): 1–44.

8. In each year of the program, more than 60 percent of the lead added to gasoline was associated with traded lead credits. See Robert Hahn and Gordon L. Hester, "Marketable Permits: Lessons for Theory and Practice," *Ecology Law Quarterly* 16 (1989): 361–406.

9. For a description of the legislation, see Brian L. Ferrall, "The Clean Air Act Amendments of 1990 and the Use of Market Forces to Control Sulfur Dioxide Emissions," *Harvard Journal on Legislation* 28 (1991): 235–52.

10. See Dallas Burtraw, "Cost Savings sans Allowance Trades? Evaluating the SO_2 Emission Trading Program to Date," Discussion Paper 95–130 (Washington, D.C.: Resources for the Future, September 1995); and Elizabeth M. Bailey, "Allowance Trading Activity and State Regulatory Rulings: Evidence from the U.S. Acid Rain Program," MIT-PAPER 96–002 WP (Cambridge: Center for Energy and Environmental Policy Research, MIT, 1996).

11. For a detailed case study of the evolution of the use of economic incentives in the SCAQMD, see NAPA, *The Environment Goes to Market: The Implementation of Economic Incentives for Pollution Control* (Washington, D.C.: NAPA, July 1994), chap. 2. Recent implementation problems with the RECLAIM program, however, illustrate a point we emphasize throughout the chapter: for a host of reasons, actual applications of market-based instruments tend not to perform up to the standards that the simplest analysis might anticipate.

12. See P. Bohm, *Deposit-Refund Systems: Theory and Applications to Environmental, Conservation, and Consumer Policy* (Baltimore: Published for Resources for the Future, in the Johns Hopkins University Press, 1981). Peter S. Menell, "Beyond the Throwaway Society: An Incentive Approach to Regulating Municipal Solid Waste," *Ecology Law Quarterly* 17, no. 4 (1990): 655–739.

13. See W. R. Z. Willey and Thomas J. Graff, "Federal Water Policy in the United States—An Agenda for Economic and Environmental Reform," *Columbia Journal of Environmental Law* 1988: 349–51.

14. See James T. Hamilton, "Pollution as News: Media and Stock Market Reactions

to the Toxics Release Inventory Data," *Journal of Environmental Economics and Management* 28 (1995): 98–113; and EPA, *1994 Toxic Release Inventory: Public Data Release* (Washington, D.C.: EPA, January 1996).

15. See Daniel C. Esty, *Greening the GATT* (Washington, D.C.: Institute for International Economics, 1994).

16. There have been some genuine enthusiasts for market mechanisms. See Stephan Schmidheiny with the Business Council for Sustainable Development, *Changing Course: A Global Business Perspective on Development and the Environment* (Cambridge: MIT Press, 1992).

17. See Nathaniel O. Keohane, Richard L. Revesz, and Robert Stavins, "The Positive Political Economy of Instrument Choice in Environmental Policy," paper presented at the Allied Social Science Associations meeting, New Orleans, Jan. 4–6, 1997.

18. "Economic rent" is that part of an indvidual's or firm's income which is in excess of the minimum amount necessary to keep that person or firm in its given occupation. It is sometimes thought of as above-normal profits, such as those that accrue to a monopolist or the owner of a scarce resource.

19. See Don Fullerton and Thomas C. Kinnaman, "Household Responses to Pricing Garbage by the Bag," *American Economic Review* 86 (1996): 971–84.

20. See Lisa A. Skumatz, "Beyond Case Studies: Quantitative Effects of Recycling and Variable Rates Programs," *Resource Recycling*, September 1996: 62–68.

21. See William Fulton, "The Big Green Bazaar," *Governing Magazine* (June 1996): 38.

22. See Hilary A. Sigman, "A Comparison of Policies for Lead Recycling," *RAND Journal of Economics* 26 (1995): 452–78.

e i g h t

Privately Financed Sustainable Development

Stephan Schmidheiny and Bradford Gentry

A small but growing group of developing countries is wrestling with a welcome but difficult new problem: how to manage major inflows of private capital from the industrial world. Environmental policymakers are likewise struggling to respond to the shift from foreign aid to private capital as the engine of sustainable development. These trends have important implications for U.S. competitiveness and environmental policy.

The concept of sustainable development—meeting needs today without stealing or wasting assets required by future generations—makes development efforts more complex and often more expensive in both the developing and industrialized worlds. In developing nations, it means setting up legal and economic frameworks that provide young and rapidly growing populations greater access to jobs, education, housing, energy, water, food, sewage treatment, and transport, without polluting air, soil, and water, wiping out forests and genetic resources, or accelerating global climate change.

Few officials in developing countries deny that sustainable development is the ideal. But since most poor countries cannot provide basic infrastructure, they are unlikely to be concerned about protecting plant species that future generations may or may not need. Moreover, some countries question why they should be asked to leave their natural resources undeveloped or take extraordinary steps to reduce pollution when many global environmental problems—depletion of the ozone layer, declining biodiversity—began during the industrialization of the now industrialized world. So they expect some help in protecting the environment, especially on the global problems.

As a result, the development debate has traditionally focused on the amounts and impacts of official development assistance, or foreign aid.

Developing-country governments and international agencies have assumed that additional resources would have to arrive in the form of low-cost loans or outright grants from industrial countries or the multilateral agencies they fund. As recently as the 1992 Rio Earth Summit, it was estimated that between 1993 and 2000 the developing world would have to spend $600 billion each year in order to put its economies on the path to sustainable development.[1] The summit included a call for at least $125 billion of this sum to come each year in the form of official development assistance.

However, that appeal for increased foreign aid has largely been ignored. Since 1992, official development aid has decreased rather than been augmented. According to the World Bank, from 1990 to 1995 official aid averaged only $57 billion annually, even including the extraordinary amounts lent to Mexico in 1995. Given the political atmosphere in the United States and elsewhere in the industrial world, aid levels are likely to continue to decline.

Shifting to Private Finance

Notably, a number of countries in the developing world are receiving much of the additional foreign capital they need—but from private investors rather than government agencies. International private investment in developing countries virtually quadrupled from $44 billion in 1990 to $167 billion in 1995. In stark contrast to aid flows, annual private capital flows to developing countries have exceeded the Earth Summit's goal of $125 billion in new and additional resources each year since 1992 (fig. 8.1).

The shift from foreign aid to private finance has occurred for a number of reasons: a growing acceptance by governments of market- rather than state-directed economics; the resulting transfer to the private sector of public enterprises and services; a growing recognition of the benefits of liberalized trade and investment regulation; and internationalization of the world's financial community, now little troubled by national boundaries in its search for profitable investments (see chapter 6). The link with trade is strong, as much of this private investment is used to establish bases either for exports or, increasingly, for serving local or regional markets directly rather than through exports.

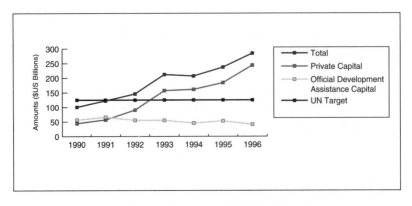

8.1 Sources of Capital to Developing Countries (U. N., World Bank)

Slightly more than half of this private capital goes into the financing of factories and other businesses as foreign direct investment. Such FDI includes manufacturing plants established and owned (in whole or in part) by foreign corporations, such as General Electric's operations in China and India; investments in once government-owned but now privatized ventures like the SIDER-MEX steel companies in Mexico; and participation by private groups in what once would have been government-funded infrastructure projects, such as the Paiton I coal-fired electric power generating plant in Indonesia. Another third of the private capital flow is debt—either direct loans from commercial banks, such as to the Sual power station in the Philippines, or proceeds from bonds sold to private investors in the international capital markets, like the U.S. bond offering for the Leyte-Luzon geothermal project, also in the Philippines. The balance of the private capital flow consists of portfolio investments in securities issued by companies in developing countries that are either traded on the local stock exchange or, in cases such as YPF (the privatized Argentinean gas company), on the U.S. or other foreign exchange.

One major concern has been whether private investors will stay the course. Can developing countries rely on a continued flow of such funds? So far, the answer is yes, provided a country keeps its side of the bargain by maintaining a free and open market for investment and goods. In the winter of 1994–95, the Mexican economic crisis did temporarily shake the confidence of portfolio investors. Having increased their holdings in developing countries tenfold between 1990 and 1993, by 1995 portfolio investors had cut their new investments by over 50 percent from the 1993 high. Direct investors and lenders, who can't take their factories elsewhere nor call their loans overnight, generally stayed the course. As a result, despite the Mexican crisis, the amount of private money flowing to the developing world in 1995 increased by 5 percent. The World Bank now predicts that foreign direct investment in developing countries will continue to grow at 7 to 10 percent per year over the next decade.[2]

Another major concern is that the flow of private capital is benefiting only a few developing countries, and the recipients tend to be the ones who have already demonstrated an ability to help themselves. Since 1990, most of it has gone to only a few countries: Argentina, Brazil, China, Hungary, India, Indonesia, South Korea, Malaysia, Mexico, Russia, Thailand, and Turkey. In fact, some of these countries, notably Malaysia and Thailand, have been so successful at attracting international capital that they are considering restricting foreign investment in order to keep their economies in balance. Although these countries represent a large percentage of the world's total population and several have great stores of biodiversity, they are only twelve in number. Other developing nations are attempting to increase their share of private flows by eliminating investment restrictions and taking other steps to encourage private investors. Countries that are not successful—such as those in the poor areas of sub-Saharan Africa or parts of South Asia—are in serious danger of being left further and further behind the rest of the world. Continued reductions in foreign aid only increase their plight as the global economy moves from a split between the "developed" and the "developing" worlds to a wider spectrum of levels of development.

Clearing the Air

Many environmentalists (as well as economists and development specialists) are struggling to understand the implications of this shift from public to private funding. Who is in charge? What are the environmental impacts? What are the pressure points for improving environmental performance? What are the roles for public and private actors? Although much remains to be done, recent research suggests answers to some of these questions.[3]

Clearly, increased private investment has led to some new environmental damage. Combining the developing world's traditional underenforcement of existing environmental regulations with new projects created in the expectation of profit can lead quickly to environmental problems. Developments in and around the manufacturing "free zones," or *maquiladoras*, in Mexico have stressed water resources along the border. Expansions of the acreage devoted to monocrop agriculture in Costa Rica (bananas) and Brazil (soybeans) have substantially altered local ecosystems and increased chemical loads.

Improvements in environmental performance can also accompany private capital flows, however, through the introduction of new investment funds and management. Privatization of the municipal water and sewerage system in Buenos Aires, with the help of foreign companies and capital, resulted in the provision of more and better water at lower rates, as well as the identification of hundreds of industrial facilities—once overlooked or tolerated by government employees—that had been failing to meet waste water discharge requirements.

Altos Hornos de Mexico S.A. (AHMSA), owner of the largest integrated steel plant in Mexico, provides another example of environmental progress through privatization. In 1991, prior to the sale of the company to a consortium of Mexican and Dutch investors, the government investigated and identified environmental problems at the plant. It then incorporated a schedule of remedial actions into the privatization contracts. As a result of privatization, the open-hearth furnace has been closed, cracks in the coke ovens have been sealed, and acidic wastewater discharges have been eliminated. With help from its Dutch investor, the company has received international and European certification of its manufacturing standards, increased its exports to Europe, and installed a state-of-the-art environmental management system.

Tapping the environmental experience of international companies in their home markets is another important environmental benefit of the new private capital flows—as shown with the Dutch investors in the case of AHMSA. In response to intense environmental pressures in the industrial world, many corporations have adopted extensive programs for managing the commercial impacts of environmental issues on their businesses. As these companies invest in the developing world, they bring with them modern, often cleaner, technology, proactive environmental management systems, and environmental training programs.

They also bring a need—and thus help create a local market—for people trained in environmental issues and for suppliers of environmental services and equipment (such as wastewater and solid waste treatment facilities). This need can bring benefits that extend well beyond the multinationals themselves. In Juarez, Mexico, for example, a number of international companies including Ford, General Motors, Johnson and Johnson, Philips, and United Technologies are working with local and national officials to promote the construction of regional wastewater treatment plants. Elsewhere, foreign investors are actually constructing and operating environmental infrastructure projects. French-based Lyonnaise des Eaux is providing water in Argentina. NorthWest Water, a British company, is providing sewerage services in Malaysia. And a U.S. company, WMX Technologies, is operating a hazardous waste treatment facility in Hong Kong.

Changing the Government's Role

Back in the days when foreign aid was seen as the primary engine of development, governments were held responsible for the environmental impacts—good and bad—of those aid flows. Although neither donor nor recipient governments were viewed as particularly good at maximizing the environmental benefits of aid, at least their responsibilities for attempting to do so were clear.

The nature of private capital flows makes responsibilities less obvious. The environmental challenge to developing-country governments is to protect against environmental risks while keeping the process as open as possible and capturing the benefits. Doing so is consistent with the changed role for governments in many developing countries—moving from the provider of goods and services to the enabler and overseer of private activity.

In order to make the best of these new capital flows, new rules need to be adopted and new skills acquired by many developing-country governments. First, governments must establish and maintain the basic framework necessary to support any private investment, foreign or local: macroeconomic stability; secure property rights; convertible currency; ability to repatriate profits; and fair and effective dispute-resolution mechanisms. Second, from a purely environmental point of view, they must set the prices to be paid for using environmental resources. This can be done with any of a wide range of policy tools—from emission limits to pollution taxes—to force the internalization of environmental costs. Without this framework, market forces alone create incentives to spill pollution harms onto others.

Developing-country governments will only be willing to set and enforce such environmental prices if they believe their competitiveness will not suffer. If they think they are limiting economic growth or discouraging new investment by forcing local or foreign-owned companies to absorb costs not borne by competitors in other countries, they will not follow through on environmental issues. This is the traditional position in much of the developing world, leading to lots of environmental requirements on the books, few of which are enforced.

Even in the face of these legitimate competitiveness concerns, developing countries are finding that it is almost always in their best interest to implement stronger environmental policies. First and most important, pollution and other environmental ills damage the health, way of life, and productivity of their citizens, denying them the benefits of any economic gains. On a narrower level, clear and consistently applied environmental policies meet the private investors' need for predictability in government requirements and simply make the country a more attractive location for investment. In fact, evidence is accumulating that higher environmental standards do not drive away private investors. When most investors are considering the location for a particular investment, environmental costs are usually far outweighed by considerations such as labor costs and access to markets. Finally, consumers in the industrial world, particularly northern Europe and, to a much lesser extent, the United States, are beginning to insist on respect for the environment in their buying decisions, setting requirements for minimum standards of environmental performance and "environmentally responsible" products regardless of the country of origin.

Developing countries' efforts to address the environmental impacts of development can be helped by governments of the countries from which the capital comes and by international development banks. Both can provide technical and financial support to integrate private investment and environmental improvement. They can also take environmental impacts into consideration when deciding on the provision of aid or direct financial support to private investments. Capital-source countries can also encourage purchases of sustainably produced goods and services from developing countries in their domestic markets. Many such efforts are in the best commercial interests of industrial nations since they tend to broaden business opportunities. More important, the easy and large gains in protecting the environment have already been achieved in much of the industrial world. Reducing pollution in the developing world can be the most cost-effective way to attack global problems such as emissions of greenhouse gases.

Influencing Investors' Decisions

In order to harness the power of private capital flows to the goal of sustainable development, policymakers need to focus on how to influence investor decision-making. Private investors will not take the environmental impact of an investment into consideration unless there is a clear economic gain or loss associated with it. The goal is to design policy tools that have a predictable and concrete impact on investors' profits. In order to change investor behavior, the question "what's in it for me?" needs to be answered in terms relevant to their overriding business goal—making profitable investments as efficiently as possible.

An investor is driven by two key factors: risk and reward—the predictability and acceptability of the expected return on any potential investment. In many industrial countries, risk and reward have been influenced in favor of environmental issues by imposing costs such as fines for exceeding wastewater discharge limits or forcing expensive cleanup of contaminated sites.

In the developing world, however, even when regulations exist, enforcement is frequently lax or avoided through bribes or political pressure. When a credible threat of environmental enforcement does not exist, pressure from parent companies, export customers, multilateral lending agencies, and even local or international groups can help fill the gap. The beneficial links between international private investment and the environment manifest themselves in a number of forms, including:

Parent company policies. Baxter Healthcare S.A., a subsidiary of Baxter International, produces medical equipment in one of Costa Rica's industrial free zones. Although there have been many criticisms of lax environmental enforcement within the free zones, Baxter's facility has implemented a sophisticated environmental management program, including a 70 percent reduction in landfilled wastes through reuse and recycling programs. The company has also initiated environmental qualification programs for their local suppliers and introduced an environmental education program in the local primary schools.

Export customer pressure. The Brazilian soybean industry has targeted environmentally motivated consumers, including those in northern Europe, as part of its effort to diversify soybean markets. Working with the government's agricultural research corporation, the industry has embarked upon an extensive effort both to decrease the environmental impacts of soybean production (through increases in productivity in dry soils and bred-in resistance to pests) and to diversify the final uses of soybeans to include more environmentally-friendly products (such as inks and bio-diesel fuel).

Lending agency pressure. Multilateral and government agencies such as the World Bank and the U.S. Export-Import Bank now require that environmental issues be considered in any project they support.[4] A requirement for high environmental performance was, for example, one of the conditions of involvement of the World Bank Group in the privatizations of Aguas Argentinas and AHMSA described above. Recently, national development banks in a number of countries have begun to implement similar policies. For Brazil's national development bank (Banco Nacional de Desenvolvimento Econômico e Social, or BNDES), this takes the form of a "Green Protocol" that encourages federal public lending to environmentally friendly projects.

Local and international pressure. Privatization of the biggest state-owned mining company in Peru (Centromin Peru), was sent back to the drawing board as a result of articles in the local and international press on the environmental issues posed. Because of the company's high levels of proven and expected mineral reserves, a total of twenty-eight companies, including firms from Canada, England, Japan, and China, signed up to participate in the bidding. However, during the first call for bids in April 1994, none of the companies submitted a proposal and the auction was declared a failure. The government later discovered that articles had run in both the local press and *Newsweek* magazine on the environmental damage being caused by the company and the government's failure to clarify responsibility for addressing those issues as part of the privatization.

The opportunities for policymakers to affect the environmental consequences of private investment vary according to the type of investment being made. In general, they depend on: (i) the time frame within which an acceptable return is likely to be received; (ii) the directness of the investment's potential impact on the environment; (iii) the level of return that is acceptable for that type of investment; and (iv) factors specific to the particular investment, including its location, sector, and size.

Foreign direct investments offer the greatest opportunities for environmental policymakers to influence investors' decisionmaking. The investor's time frame is generally medium- to long-term. Many projects pose direct risks to the environment through their potential for increased resource use, air emissions, water discharges, and waste generation. At the same time, many can be effective vehicles for the transfer of new environmental technologies and management systems. In fact, in response to commercial risks, environmental issues are already given careful consideration by many investors during their decisionmaking on any particular FDI project.

Fewer leverage points exist at the other extreme—portfolio equity flows. Once shares are listed, portfolio investments tend to involve large pools of capital (such as mutual or pension funds) in search of the greatest returns in the short to medium term. Given their short-term focus, they are the most volatile. Moreover, unless the shares are in a company that offers environmental goods and services or faces catastrophic environmental costs, environmental considerations are unlikely to be material commercial factors in the short time frames of concern. The result is less need to understand and address environmental issues when making investment decisions.

The U.S. Government Role

Although developing nations bear the ultimate responsibility for mitigating the environmental impact of their economic growth, the United States (as one of the most important sources of private and public capital) can and should play an important supporting role, as should other industrial countries. This commitment should include efforts to help build markets for profitable investments resulting in environmental improvements, to increase the quality of information flows on such investments, and to provide direct financial support for investments that are particularly attractive from a policy point of view.

Supporting Local Markets

The U.S. government should take a number of steps to help developing countries establish local incentives for reflecting environmental considerations in private investments.

Provide technical support for developing-country efforts to increase the importance of environmental factors in local investment decisions. Developing countries should lead investors to consider the importance of environmental factors in their local investments by: (i) consistently enforcing locally adopted, performance-based environmental requirements, (ii) establishing clear environmental liability systems, (iii) getting the price right by reducing or eliminating subsidies (on energy, water, or agricultural products) or increasing fees on emissions, (iv) increasing the public disclosure of environmental information by companies, (v) applying environmental screens to their national development assistance efforts (such as the Green Protocol in Brazil), and (vi) implementing other types of higher-value-for-money environmental policies, including many described elsewhere in this book. The U.S. government should support these efforts through its technical assistance programs, particularly as part of integrated packages for increasing private investment in recipient countries.

Integrate environmental issues into harmonized investment frameworks. The twenty-seven OECD countries are currently negotiating a Multilateral Agreement on Investments (MAI) aimed at liberalizing treatment of foreign direct investments, ensuring that the basic conditions for private investment are in place, and protecting foreign investors' rights. Efforts are underway to extend the agreement to non-OECD countries. The U.S. government should press to have environmental considerations reflected in the agreement. Doing so might lead to support from a surprising quarter: U.S.-based multinational manufacturing corporations. These firms are increasingly concerned not only about the impact of environmental issues on their operations outside of the United States, but also about the possibility that environmental regulations in developing countries will be unequally enforced to their disadvantage. Should negotiations on the MAI be followed by actions within the World Trade Organization or other international agencies to develop harmonized standards for FDI, environmental factors should be incorporated into those initiatives as well.[5]

Partner with private companies on environmental education programs. As noted, many international companies have strong commercial incentives to adopt global environmental management systems, including training programs for employees and contractors. It is also in their commercial interest to increase the general level of environmental awareness and performance in countries so as not to be too far in front of local regulators and companies. The United States should help developing-country agencies take advantage of this by joining multinational companies in implementing educational programs for local businesses and government agencies.[6] Among the most useful are programs for increasing the predictability of the impact of environmental risks on particular investments and reducing operating costs or raising revenues through improved environmental performance. For example, DuPont has committed itself to drive toward zero emissions, not to comply with U.S. law but to increase the efficiency of its raw-material usage and competitiveness.

Support the incorporation of environmental factors into national income accounts. One of the most widely used indicators of the success of a country's economic development efforts is growth in its national income accounts, such as gross national product. Changes in the environmental conditions in the country are, however, currently excluded from such calculations. For example, the *Exxon Valdez* oil spill resulted in an increase in U.S. GNP by only counting the revenues generated by the clean-up activities and not the damage to ocean resources. Supporting the efforts to incorporate environmental factors into national income accounts will help build local political support for integrating environmental factors into national development plans.

Reach a multilateral agreement linking foreign aid to compliance with baseline environmental standards. Conditioning the receipt of foreign aid on conformance with baseline environmental standards will go a long way toward increasing the environmental component of private sector flows. Although the international development banks and U.S. agencies such as the Export-Import Bank apply such environmental screens to their support of private investments, export credit agencies in most other countries do not. Efforts to require that similar environmental screens be applied by all OECD member countries should be intensified. In addition, where a recipient country government ignores the international environmental impacts of national development programs—such as the rapid expansion, often without emission controls, of coal-fired power stations in China—coordinated international action to establish and apply baseline environmental standards more broadly should also be considered.[7]

Encouraging International Investors

A prime goal for U.S. policymakers should be to help attract private investors from industrial countries to projects that contribute to sustainable development. In addition to the obvious environmental benefits of this approach, recipient countries are more likely to reflect environmental issues in their own development strategies if doing so leads to immediate financial benefits, not just to improved environmental conditions. There are several ways in which U.S. and other capital source countries can help do this.

Investment products secured by environmentally responsible assets in developing countries. As baby boomers and their mutual fund managers seek to diversify their investments and take advantage of the higher growth rates in the developing world, the demand for opportunities in emerging markets has increased. This creates a market for sales in the United States of financial products that are both profitable and secured by environmentally beneficial assets or operations in recipient countries. For example, in 1995, almost $300 million in bonds were sold to U.S. institutional investors as part of the funding of the Casecnan project, a combined irrigation and hydroelectric power facility in the Philippines. Other developers are planning to tap the U.S. markets to finance projects ranging from large wind power facilities to bundled or securitized interests in a number of smaller projects (such as leases of dispersed household photovoltaic units in several Asian and Latin American countries). The U.S. government can help support such investments in a variety of ways, including offering political risk insurance, regulating the disclosure of material investment information that includes environmental aspects, helping to disseminate information on such opportunities, and purchasing such securities for its employees' pensions.

Sustainably produced goods from developing countries. The biggest— and most difficult to capture—opportunity lies in encouraging U.S. consumers to buy environmentally sound commodities from developing countries, such as sustainable forest products or responsibly extracted minerals. Sales of such products could be encouraged through government support for the establishment of product certification programs and their use in government procurement. The Forest Stewardship Council, for example, is developing standards for sustainable forest products. The government could also help by reducing subsidies to nonsustainable U.S. commodity producers, such as certain sugar operations. However, for this to work, policymakers will need to overcome recipient-country concerns over disguised trade barriers, as well as the traditional push by many U.S. interests for the government and consumers to "buy American" (narrowly defined).

Emission and other environmental credits from investments in developing countries. In general, it costs less to improve the environmental performance of manufacturing operations, power plants, and similar facilities in developing countries than it does to achieve similar improvements in industrial countries. For example, substantially increasing the energy efficiency of a steel complex in eastern Europe today, thus reducing emissions, is much less expensive than doing so in Japan or the United States. In some cases, it is possible to exploit this to everyone's benefit by shifting an investment from an industrial to a developing county using domestic tax or regulatory incentives. Under an agreement between the U.S. and Costa Rican governments, for example, Northeast Utilities (the largest electricity-generating company in New England) will receive credit against its U.S. greenhouse gas emissions reduction targets for its subsidiary's construction of a twenty-megawatt wind power facility in Costa Rica. Under the deal, all three parties benefit. The utility company profits from its Costa Rican investment and gets lower-cost emission reductions than might have been possible in the United States. The United States promotes the global reduction of greenhouse gas emissions in a cost-effective manner. And Costa Rica gets electricity from a clean wind power project.

Expanding Information Flows

Nothing is more important to private investors than information. And in much of the developing world, nothing is more difficult to obtain. Profound differences in accounting systems and willingness to disclose financial data have long made it difficult to compare competing investment opportunities in different countries.

For the environmentally concerned investor, the problem is magnified by a near universal lack of quantified data on the financial impacts of environmental performance. With the exception of the disclosure requirements for environmental risks imposed by the U.S. Securities and Exchange Commission, little information on the financial impacts of environmental factors is required to be disclosed by companies in either the industrialized or developing world.

Given the importance of full information to investors, mainstreaming the positive or negative financial impacts of environmental considerations into traditional financial market information channels is the critical long-term goal. The U.S. and other capital-source–country governments could help expedite this process by supporting the integration of environmental factors into harmonized international standards for accounting and stock exchange reporting. Although much remains to be done to standardize these rules, efforts should be made to include environmental information in project or corporate reporting. The U.S. experience with securities regulation requirements for disclosure of environmental risk information, though not perfect, provides valuable lessons for these international efforts.

However, even if such harmonized reporting standards were in place, major communication gaps would still exist as a result of the different languages used by the environmental and financial communities, as well as the lack of accepted mechanisms for quantifying the financial implications of environmental risks and opportunities. For example, while evidence is mounting at companies like Dow that eco-efficient behavior[8] improves their competitiveness (often by reducing production inputs, hence production costs), these same companies have yet to find those benefits translated into lower costs of capital. Although the public sector cannot cure these private sector difficulties, the U.S. government should be able to play a catalytic role by supporting research and networking both within the financial community and between the financial community and its customers in order to overcome these language and valuation issues.

Ultimately, private investors will collect the information they believe necessary to justify any particular investment—or they will look at other deals. U.S. government agencies can also help prevent investors from overlooking suitable projects by making basic information on the environmental aspects of investments in developing countries readily available.

Working Together

Even with the ascendancy of private capital, direct public support for particularly attractive projects will continue to make sense. Not all socially and environmentally desirable projects will initially be able to attract private capital. Some may not generate sufficient revenues quickly enough. Others may be in countries or sectors considered too risky by the financial community. Still others may not be big or profitable enough. Governments can improve the attractiveness of these investment opportunities and clear the way for private capital through a wide range of financing structures. The options include projects fully funded by the government but managed by private operators, mixed public and private sector funding, and government "standby" protection for specified risks.

For governments to help pave the way for private capital through direct financial support, representatives of both government and business will have to overcome their mutual mistrust and different working practices. Although public officials speak of the need to leverage government assistance through partnerships with the private sector, many balk at the notion of using public money to help companies make a profit. At the same time, many people in business, particularly in the financial sector, view attempts to work with governments as a complete waste of time given the delays, politics, level of disclosure, and unpredictability involved. Such barriers to an expanded range of public-private partnerships can, however, be overcome. The private sector, for instance, regularly finds it possible to take advantage of such "traditional" public sector financial support as export credit and promotion programs or political risk insurance.

To the extent the U.S. government provides financial assistance to particularly attractive environmental projects, it should focus on demonstration projects that go beyond the technical performance of new equipment to proving the commercial value and replicability of the particular technologies and financing structures. It should also concentrate grants or other concessional aid on ways to reduce the up-front transaction costs of projects, rather than a continuing commitment to cover shortfalls in operating revenues. Concern over the use of public funds to help generate private profits could be lessened by opportunities for government agencies to earn a profit when the project becomes a commercial success.

The government could also expand its risk-mitigation tools such as political risk insurance to include a broader suite of techniques for overcoming financing obstacles such as the project's location or the time required for it to become profitable. For example, if bank financing is available only for the first eight years of a fifteen-year project, a government entity could offer a standby guarantee to provide financing after year eight in the event that private capital for refinancing could not be found at that time. Such mechanisms would be particularly useful as part of efforts to bundle a number of small environmental projects (such as solar photovoltaic systems) into a financial package large enough to interest international investors by providing extra layers of certainty that investors will be repaid.

It can also be helpful to create strategic alliances with a broader range of parties seeking to promote such investments, including particular countries or subnational units (such as the joint implementation projects in Costa Rica), the private managers of environmental investment funds, local community development organizations, nongovernmental organizations, and small-business associations. Small and medium-sized enterprises should be particular targets for support, as a result of their traditional lack of access to capital and information, as well as the substantial levels of environmental damage they cause in many developing countries.

Steps on the Path Toward Sustainable Development

Although all of these tools will help increase the environmental benefits of private investment flows to developing countries, they will not—standing alone—lead us to a sustainable future. First, the social aspects of sustainable development must also be reflected in private investments through means still being developed. Even more fundamental, our collective understanding of what it takes to build a sustainable future—integrating economic, environmental, and social considerations—is extremely limited in both industrial and developing countries.

Nevertheless, the growth in private capital flows to the developing world holds great promise for improved environmental performance. The United States, as one of the world's largest foreign investors and recipients of overseas capital, has a leading role to play in developing mechanisms to ensure that this promise is realized. It is now time for the United States and individual developing countries to work together with investors to capture the opportunity for privately financed sustainable development.

Notes

1. United Nations Conference on Environment and Development, *Agenda 21* (Washington, D.C., 1992), chap. 33.

2. World Bank, *World Debt Tables* (New York: World Bank, 1996).

3. Bradford S. Gentry, ed., *Private Capital Flows and the Environment: Lessons from Latin America* (forthcoming).

4. Export-Import Bank of the United States, *Environmental Procedures and Guidelines* (Washington, D.C.: Export-Import Bank of the United States, February 1995).

5. E. Graham, *Global Corporations and National Governments* (Washington, D.C.: Institute for International Economics, 1996).

6. The U.S. Environmental Training Institute presents a good model for the type of program required.

7. Daniel C. Esty and Robert Mendelsohn, "Powering China" (unpublished study, Yale Center for Environmental Law and Policy).

8. Stephan Schmidheiny with the Business Council for Sustainable Development, *Changing Course: A Global Business Perspective on Development and the Environment* (Cambridge: MIT Press, 1992).

nine

Technology Innovation and
Environmental Progress

John T. Preston

Since the industrial revolution, two major trends have placed stress on our environment. First, the world's population has grown to the point where some of our wastes can no longer be disposed of simply and safely. Second, technology has given us the tools—from cars to mines to harvesters—to exploit the natural world and thereby to improve our lives. But these same technologies have side effects—pollution, habitat destruction, land degradation—that can endanger our quality of life. Despite technology's mixed past, it represents perhaps the most promising avenue for an improved environment in the future.

Technological advances make emissions controls cheaper and more effective. Innovation can also help us to identify new, less polluting ways to manufacture, distribute, and consume products. Even technologies that seem entirely unrelated to the environment, like the availability of information over the Internet, offer the potential for significant environmental gains. In fact, a more technologically sophisticated and information-intensive society offers the promise of becoming less dependent on the consumption of physical materials and less reliant on polluting activities. Thus, the sooner we innovate, the sooner we can reduce pollution.

A central focus of next-generation policymaking must therefore be on encouraging innovation and paving the way for the rapid adoption of environmentally beneficial technologies. Technological innovation, the commercially successful application of new ideas, can be accelerated and pointed in a general direction through government intervention, industrial research and development, and entrepreneurial zeal. The case of automobile emissions provides an example of how these three technology drivers make themselves felt. Tailpipe emission standards were intentionally set by Congress at levels that were known to be beyond the capacity of the then-available technology. These performance-based regulations led to the commercial application of catalytic converter technology and to the phase-out of leaded gasoline. Petroleum companies have now adopted research results from automobile companies to "reformulate" gasoline, demonstrating that internal combustion can be friendlier to the environment than it has been in the past. Finally, the market for alternative-fuel cars, driven largely by regulation, has created opportunities for entrepreneurs.[1] Today, a hotly contested race is under way to produce commercially viable electric cars and other low- or zero-emission vehicles.

We have strong evidence of why optimism about technology is well placed. During the last decade, the chemical industry has reduced its emissions of noxious materials by 50 percent while doubling its output. Similar but less dramatic results have occurred in nearly every industry, ranging from paper to aerospace, and even less obvious industries like banking and insurance. The impetus to incorporate new technologies that can improve the environment has come from many directions.

Some improvements are driven simply by applying good engineering practices. More than 10 percent of the methane transported in Russian pipelines is, for example, leaked into the atmosphere, where it contributes to the greenhouse effect and global warming. In western Europe and the United States, the comparable loss is about 1 percent. In addition to the obvious environmental gain from adopting technology to detect and fix leaks, there is a clear economic one—10 percent greater output without digging new wells.

Improvements in business practices have also led to an understanding of why updating environmental technologies can be very important. Managers at an Amoco oil refinery, for example, originally estimated that environmental considerations amounted to 3 percent of plant operating costs. A careful study using better cost-accounting methods revealed that environmental factors were actually contributing more than 20 percent of the costs of running the refinery.[2] We often find that when all factors are considered, the cost of pollution is much higher than originally estimated. This means that investments in technology to reduce pollution can provide a return to the company much more quickly than was originally expected, that is, they have a shorter payback period.

Not only businesses but also governments are starting to recognize the true costs of environmental damage and the need for faster adoption of technological solutions. Recently, a study in the United States determined that air pollution results in more than sixty thousand deaths per year—more people than are killed in automobile accidents each year, and more Americans than were killed during the entire Vietnam War. The government of Chile has quantified the health costs incurred because of the polluted air in Santiago and found that during the winter nearly thirty-five hundred children are treated each day for respiratory ailments. By recognizing the billions of dollars spent on treating people for the effects of pollution, Chile is realizing the economic value of investing in environmental improvement.

Finally, pressure from environmental groups, customers, and regulators has stimulated much change. Corporate leaders want to preserve their image within their communities and maintain good customer and supplier relations. When those relations are jeopardized, change can occur quite rapidly—in essence, shortening the payback period. For example, when Mitsubishi products were boycotted because a subsidiary was using practices that were destroying rain forests, Mitsubishi promptly intervened, moving its subsidiary to adopt technologies to reduce its environmental impact.[3]

Range of Environmental Technology

The National Science and Technology Council has subdivided environmental technologies into four categories. Monitoring and assessment technologies establish and track the condition of the environment and harmful releases to the environment. Avoidance technologies minimize environmental damage through process innovation and pollution prevention. Control technologies prevent hazardous substances from entering the environment. Remediation and restoration technologies improve the environment after it has received human or natural harm.

Monitoring and assessment technologies. As a result of engineering and scientific advances, we have seen steady improvement in our ability to measure smaller amounts of emissions and to tie them to their sources. We are also learning more about the health effects caused by environmental contaminants. For example, technology is being developed that will enable us to pinpoint the cause of a mutation in human cells. Thus, for the first time, we can quantify the environmental causes of cancer. Monitoring technologies are also being used to control traffic jams and to minimize chemical inputs in food production.

Avoidance technologies. These technologies range from modeling to source reduction to materials substitution. A common use of modeling in the design process is to predict the cost, strength, quality, safety, durability, and manufacturability of a new product. There is an interesting environmental impact associated with this kind of modeling: prototypes do not all have to be built and tested, which saves not only money but also materials and energy.[4] Now, however, organizations are starting to use modeling specifically to predict environmental impacts. The U.S. Navy is modeling the environmental impacts of new aircraft designs, allowing refinements to the plans to be made more cheaply at an earlier stage in the production process. The use of such modeling will increase as environmental costs are applied more fully throughout the expected life-cycle of a product from production to consumption to disposal.

Smart designs of new products or processes that reduce energy and materials use yield immediate positive results. For example, new semiconductor chips are faster and more powerful yet use less energy. Cars have become both safer and lighter using fewer resources, including the substitution of new composite materials for steel. We are also seeing an increasing number of cases where more benign materials are being substituted for hazardous ones. The use of lithium ion in place of nickel cadmium in batteries, for example, reduces the problems associated with the disposal of batteries containing heavy metals that contaminate landfills. From 1980 to 1990, the impact of technology-induced pollution reduction has been dramatic: ambient lead levels have been cut by 88 percent (primarily due to the removal of lead in gasoline), particulate matter by 21 percent, carbon monoxide by 31 percent, volatile organic compounds by 15 percent, and sulfur oxides by 35 percent.

Control technologies. End-of-the-pipe technologies for controlling air, water, and waste make up the largest subgroup of environmental technologies measured by current market revenues. One of the exciting trends over the last ten years has been the conversion of what were previously wastes into usable products. Old tires are being recycled into new uses, such as marine products.

One company is injecting hazardous materials into a molten metal bath, catalytically breaking the waste into its elements, which are then brought back into commerce as products. Another company has developed a method for reusing asphalt for highways. This approach not only permits the reuse of asphalt but also lowers the amount of new oil needed to make a one mile lane of highway from more than a hundred barrels to only three.

Remediation and restoration. Finally, technology is being used to restore contaminated sites. One technique under investigation is the use of osmosis to "drive" wastes into a smaller area or to prevent wastes from leaching into water supplies. In still other experiments, supercritical water is being used to break down organic wastes into carbon dioxide and a benign residue. And plants are being engineered to absorb heavy metals and radioactive isotopes into the plant cellulose. When these plants are harvested, the contamination is now concentrated in a small amount of plant biomass rather than a large amount of soil.

The Problems

Despite all the incredible inventiveness in our companies, universities, and national laboratories, technologies that could solve serious environmental problems are not being adopted as rapidly and universally as they should be. We are not commercializing environmental technologies because: (i) government subsidies create incentives for overuse of certain resources, particularly energy and water, which disadvantages innovative technologies; (ii) our current regulatory structure "undercharges" polluters and reduces the incentive for innovators to develop or adopt cleaner technologies; (iii) innovation is not funded consistently through the different stages of technology development from beginning to end; rather, there is a funding gap; and (iv) capital markets shy away from long-term risks, regulatory uncertainty, and market fragmentation.

Government subsidies. The overall effect of subsidies is to stimulate overconsumption based on old technologies. This is exactly the wrong outcome if we want to encourage innovators to solve environmental problems. With regard to water, the World Bank estimates that more than $50 billion is used worldwide to subsidize water consumption each year. In the United States alone, irrigation is subsidized at a cost of $2.5 billion a year. Water subsidies lower the cost of water supply, sanitation, and irrigation as well as making water available to a greater percentage of the population at affordable prices. However, the effects are complex. Most important, water is wasted and water-conserving technologies are not encouraged.

One particularly problematic example is the case of a contemporary entrepreneur trying to develop a technology that holds water in the topsoil. The prototype product is a spray that seals the soil and inhibits evaporation. The net effect is a two-thirds reduction in the amount of water needed. Let's assume that the cost of water is $500 per acre, for which the subsidized farmer pays $200. If the cost of the new technology to reduce the water needed by two-thirds is $200 per acre, the acreage cost will be lowered to $367 ($200 plus one-third of $500). However, the government subsidy is based on the amount of water consumed, so it also declines by two-thirds. As a result, the cost to the farmer becomes more than it was originally. Now, the developer of the new technology must convince the government to help subsidize the technology instead of water at a considerable savings for the taxpayers. Thus, this new technology faces a nearly impossible sale.

Even more money is spent subsidizing energy: the World Bank estimates the sum to exceed $450 billion. Not surprisingly, the energy sector problems thus created are larger in scale. Energy subsidies in the United States for fossil fuels and nuclear energy amount to $27 billion per year, with an additional $3 billion spent to subsidize renewable resources, predominantly hydroelectric power. Moreover, the total for fossil-fuel subsidies does not include the fact that U.S. taxes on these fuels are much lower than in the rest of the world.

The effect is that U.S. per capita consumption of fossil fuels is higher than any other country's—as are the carbon dioxide emissions and other environmental problems associated with burning fossil fuels. At the same time, fuel subsidies retard the development of competitive, less environmentally damaging technologies. Not only do new technologies need to be cheaper than fossil fuels, but they must also be cheaper than the reduced prices of fossil fuels created by the subsidies. During the oil shock in the 1970s, the U.S. government also offered subsidies for solar, wind, and other renewable energy sources, but these were abandoned, leaving fossil fuels with their government-supplied competitive advantage.

Undercharging polluters. Failing to charge polluters the full cost of their activities creates the same problems as subsidies. For example, particulate

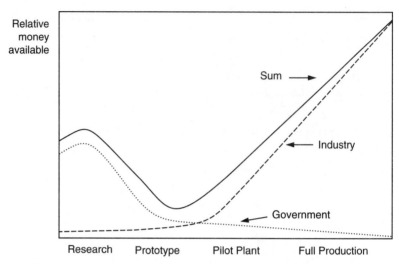

9.1 The Gap Between Public- and Private-Sector Funding of Innovative Technology

matter in the atmosphere is one cause of health problems ranging from asthma to cancer. The cost to society for the deaths, illnesses, and lost work caused by these diseases is billions of dollars per year.[5] Currently, waste generators are not assessed the full cost of disease-causing pollution. This leads to overgeneration of waste and inhibits innovative technologies. For example, if a waste generator were charged $100 per ton of particulate matter released, that polluter would be willing to pay up to $99 for a technology that eliminates the pollution. Innovation occurs where the economic advantage is the highest. Markets will not "pull" innovation unless the costs of pollution are accurately reflected in market prices.

The funding "gap." In the United States, the government funds most basic research while industry funds the later stages of scaling up technology into full production. The problem is that there is a lack of funding for demonstrating technologies at small production levels or for building the first pilot plant. Figure 9.1 shows the gap in funding that arises at the most critical point in technology commercialization—before investors are willing to make risky investments and after government officials conclude that the project is now too commercial for them to fund. The net result is a shortage of investors

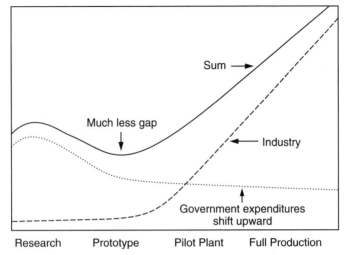

9.2 The Result of Government Investment to Fill the Funding Gap

in the middle stages of developing technologies. In the graph these stages include building prototypes and pilot plants, which constitute the first economic proof that the technology will meet market needs.

Analysis of the funding gap as it relates to environmental technologies yields some interesting conclusions using the example of hazardous wastes. It is useful to break the problem into two categories: (i) hazardous wastes generated by the U.S. government or industrial hazardous wastes that can be treated using the same technology used by the government, and (ii) all other wastes. The reason for this distinction is that the U.S. government is paying to clean up the hazardous wastes it has generated—from weapons production during the Cold War, radioactive wastes from the U.S. national laboratories, or other defense-related wastes such as nerve gas and explosives. With regard to these government-generated wastes, the federal government funds the development of remediation technologies across the stages from basic research to full-scale operations (fig. 9.2).

Some of the developments benefiting from government investment will be applicable to wastes generated by the nuclear power industry and the

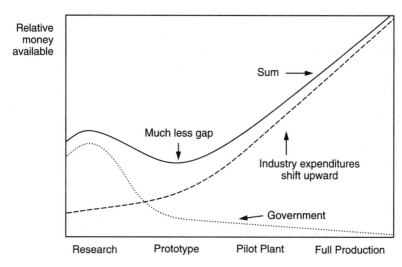

9.3 Japanese Policy Encourages Private Industry to Make Long-Term Investments to Fill the Gap

chemical industry. And such processes as classifying wastes and breaking wastes into its elements are being supported by the government and will have industrial applications. However, many environmental technologies do not address government wastes and have a much more difficult time attracting funding and getting commercialized. In *Green Gold,* Curtis Moore and Alan Miller describe company after company involved in environmental technologies that could not get sufficient U.S. financing and ended up being developed overseas.[6] The failure to get gap funding has led to European companies now having a dominant position in air scrubbing. Another example is a technology called "wet oxidation." Water that contains hazardous wastes is pressurized and heated to the supercritical state at which reactions occur that break down the wastes into benign byproducts. This technology was pioneered at the Massachusetts Institute of Technology, but because of the funding gap, the first application for the technology was in Germany. Although it is too early to tell whether this technology will also be lost to foreign competitors, clearly the indications suggest such a fate.

The funding gap is not limited to environmental technologies. U.S. inventors have pioneered many breakthrough technologies that have found more receptive investment climates abroad. Examples include liquid crystal displays led by the Japanese and hard disk drives led by Singapore. Both are multibillion-dollar product lines used in personal computers. Both were invented in the United States and are now manufactured abroad. Interestingly, these two countries have very different reasons for their success in funding through the gap. Singapore uses a centralized approach very similar to U.S. government funding of environmental technologies that treat government wastes. In contrast, the Japanese model is driven by encouraging industry to make the long-term investments needed to fill the gap (fig. 9.3).

Risk, regulatory uncertainty, and market fragmentation. The notion of the funding gap helps to explain why investments in technology in general and in environmental technology specifically are so risky. In fact, a recent paper by Joseph Romm and Charles Curtis showed that venture capital investments in environmental industries lag behind those in other industries.[7] The amount of investment in environmental technology companies in 1994 was one-tenth the amount of capital invested in biotechnology, despite the fact that U.S. environmental technology companies have twice the revenues of biotechnology companies.

Clearly, the venture capitalists are saying that the risks associated with investing in the environmental sector are high relative to the rewards. The investment community in general has shown little interest in an industry characterized by uncertainty and driven by government regulation and government spending. This in turn makes many environmental technology markets unstable. Because much environmental law is delegated to the states and each state regulates a little differently, innovators may be forced to sell to many submarkets rather than to service a unified national market. Such small markets may not justify investment in new technologies. Furthermore, environmental technologies tend to have long-term paybacks that run counter to the essentially short-term focus of the U.S. capital markets.[8]

The enormous role played by regulation in the environmental field has many other ramifications for technology and investment. Much first-generation policy focused on "technology-based regulation" that told the regulated community just what technology to choose. This was a useful way to promote the widespread use of environmental technology, rather than innovation. Such regulatory policies favor existing, older technologies over new, unproved technologies for several reasons. First, regulators often grandfather existing facilities to avoid imposing new capital burdens on companies. Second, regulators face increased risk approving a new technology over a technology that has been in practice for many years. Likewise, a company faces increased risk in trying to get a new technology approved because the regulators will be more cautious and it will require more time to educate the regulators on the merits of the new technology. This favors technology "lock-in" of the existing approach.

So, when a new technology emerges, it is immediately at a disadvantage.[9] Will the plant constructed using the new technology be able to obtain the needed permits? If approval is not obtained, the plant owner must now retrofit the plant with the old conventional technology, thus facing doubled costs for attempting to innovate. Even if a new technology is successful in obtaining the needed permits, it will almost always take longer and cost more than obtaining permits for the old technology. As a World Resources Institute report noted, "Drawn-out, expensive, inconsistent and inflexible permit procedures are great burdens imposed on innovators, particularly small, creative entrepreneurial firms."[10]

Policy responses. Technological innovation offers the possibility of improving the environment while lowering costs and boosting the U.S. economy. Next-generation policy should be oriented to:

- Level the playing field so "traditional technologies" such as fossil-fuel burning do not get hidden subsidies. Although it is attractive to propose subsidies for new technologies, this encourages the subsidy mindset of legislators and, ultimately, will be counterproductive. Once a subsidy is started, it is very difficult to remove it and today's new technology will become tomorrow's old technology.
- Investigate methods for full-cost accounting for pollution and resource use to encourage conservation and adoption of environmental technologies. Until environmental costs are gathered together within a firm, there may be insufficient rationale for "green" technological improvements to processes and products. Full-cost accounting remedies this by putting environmental costs onto the same ledger as other financial expenditures. Thus, it is an important tool to insure that business decisions will include environmental considerations. There are also implications at the level of national income accounts for different

countries. These statements now underestimate the cost of depleting natural resources.[11] Correcting this can also open up opportunities for appropriate technological adoption.

- Create incentives for investors to fund the gap or encourage government use of first-of-a-kind technologies. There are two models for providing gap funding, one involving increased government investing and the other involving increased industrial investing. In the current political climate it is unlikely that the U.S. government will increase funding for technology commercialization. In fact, we have seen a rapid reduction in federal gap-related programs such as the Advanced Technology Program run by the National Institute for Standards and Technology. Incentives for the private sector to invest in the gap could include lowering taxes for long-term investments, while raising taxes for short-term investments.

- Shift to performance-based regulations that establish a base level of pollutants and reward those who make less pollution but penalize those who exceed the base level. This will encourage adoption of new, cleaner technologies and discourage simply holding onto dirty technologies. The change from a "diffusion-based strategy" to an "innovation-based strategy"[12] is now warranted as our focus on control and remediation gives way to an era of process improvements and avoidance technologies.

Environmental regulation in the United States does not optimize the opportunities for innovation. To do so, the country needs to adopt a strategy of performance-based regulation that is flexible and that reduces the risks of innovation. Whether such regulations will accomplish the twin goals of environmental protection and innovation depends on the levels of performance chosen, how performance is measured and judged for compliance, and the degree to which we employ regulatory strategies that avoid fragmenting markets.

The U.S. Environmental Protection Agency has recognized the current legal structure's bias against new technologies and recently started several programs to foster innovation. Under Project XL, for example, the EPA is supposed to solicit proposals for new environmental technologies that go beyond compliance with current pollution control regulations. After an assessment of these technologies, some are then chosen for rapid permitting or reduced regulatory oversight in return for commitments to meet strict performance goals. This change reflects a shift away from a command-and-control strategy to a "command-and-covenant approach" to environmental regulation (see chapter 11). Whether this model will prove successful in the long term is still too early to say.[13]

Performance-based regulations give industry the opportunity to innovate to meet or exceed the required performance, focusing government's role on monitoring compliance. This, in turn, gives industry the opportunity to innovate to develop low-cost solutions. Several other programs are being started to lower risks for innovators. California has developed an innovative "technology-proving protocol" to help innovative products clear regulatory hurdles more quickly and cheaply. Regulators at both the federal and state levels are starting programs that reduce the number of approvals needed to build new facilities, including "one-stop shopping" for all permits. These innovative policies by the EPA and state agencies are starting to put old and new technologies on an equal basis.

We have an opportunity to speed the adoption of technologies that solve environmental challenges facing our children and grandchildren—at ever lower costs. Many of these already exist in our universities and laboratories; they only lack the right opportunities to reach the market. A central mission of next-generation environmental policy must be to establish market and regulatory conditions that spur innovation and entrepreneurial investment in the environmental field. Technological advances offer the promise of both environmental gains for society at large and improved competitiveness for our industries.

Notes

1. Thomas Ballantine, "Environmental Policy: The Next Generation—Technology and Innovation" (paper prepared for the Next Generation symposium on technology innovation led by John Preston at Yale University, May 1996).

2. Daryl Ditz, Janet Ranganathan, and R. Darryl Banks, eds., *Green Ledgers: Case Studies in Corporate Environmental Accounting* (Washington, D.C.: World Resources Institute, 1995).

3. Stephan Schmidheiny and Federico Zorraquin, with the World Business Council for Sustainable Development, *Financing Change: The Financial Community, Eco-efficiency, and Sustainable Development* (Cambridge: MIT Press, 1996).

4. David Rejeski, "Clean Production and the Post Command-and-Control Paradigm," in *Environmental Management Systems and Cleaner Production* (forthcoming).

5. The EPA has, in fact, launched a new regulatory effort aimed at particulates. See John H. Cushman, Jr., *New York Times*, 1 Dec. 1996: 1.

6. Curtis Moore and Alan Miller, *Green Gold: Japan, Germany, the United States, and the Race for Environmental Technology* (Boston: Beacon Press, 1994).

7. Joseph J. Romm and Charles B. Curtis, "Mideast Oil Forever?" *Atlantic Monthly,* April 1996.

8. The United States has more than $2 trillion managed by money managers. These managers can make money in two ways: by shifting wealth or by generating wealth. Many of the new short-term financial instruments that increase liquidity for the market tend to shift wealth. For example, an investor in derivatives or futures makes money if someone else loses money—a zero-sum game. Thus, if money was made, it was made by shifting wealth.

As a counterexample, look at what happens when a venture capital investment is made to create a new manufacturing business. At the end of five years, there are numerous benefits to society: the value of the new products generated, the improvement in efficiency or quality of life of the people who used those products, and the jobs created by the business.

9. Although many federal environmental laws are crafted as performance standards and not specific technology mandates, EPA guidance to the regulated community often has the effect of transforming these professional standards into technology requirements.

10. George Heaton, Robert Repetto, and Rodney Sobin, *Transforming Technology: An Agenda for Environmentally Sustainable Growth in the Twenty-first Century* (Washington, D.C.: World Resources Institute, April 1991), 24.

11. See Robert Repetto, "Accounting for Environmental Assets," *Scientific American,* June 1992: 94–100.

12. Nicholas A. Ashford, "An Innovation-Based Strategy for the Environment," in A. M. Finkel and D. Golding, eds., *Worst Things First? The Debate over Risk-Based National Environmental Priorities* (Washington, D.C.: Resources for the Future, 1994), 303–04.

13. The EPA's Project XL has run into some difficulties (see John H. Cushman, Jr., "EPA and Arizona Factory Agree on Innovative Regulatory Plan," *New York Times,* 20 Nov. 1996: A18).

Data, Risk, and Science

Foundations for Analysis

James K. Hammitt

A generation ago, determining the direction of policy to protect the environment and human health was comparatively straightforward. Many of the problems, such as dirty air and water, were easily detected, and with little existing control of environmental pollution the benefits of incremental regulation were almost certain to exceed their costs. A generation later, the appropriate direction for improving environmental regulation is less clear. Many of the easy steps have been taken. Significant point sources of pollution, such as industrial smokestacks and waste-water discharge pipes, have been controlled, leaving nonpoint sources like agricultural and urban run-off and automobile exhaust as important contributors to pollution.

As the costs of increasingly stringent environmental standards grow, it is no longer obvious that stronger regulations will necessarily improve U.S. citizens' health, much less their overall welfare. The EPA estimates that the nationwide costs of compliance with federal environmental regulations were more than $100 billion in 1990 and will exceed $150 billion in 2000 (about 2 percent of the gross domestic product).[1] Sen. Daniel Patrick Moynihan has observed that although these amounts may not be too much to spend for environmental protection, they are too much to spend unwisely.[2]

One approach to environmental hazards has been a demand for "safety" or "zero risk"—for eliminating all substances that pose risk. But as the sixteenth-century physician Paracelsus recognized, in what has become the first law of toxicology, "All substances are poisons; there is none which is not a poison. The right dose differentiates a poison and a remedy."[3] Substances necessary for life, such as oxygen, salt, and vitamin D, can be toxic or hazardous in high doses or in the wrong circumstances. Hence attempts to classify substances into categories—benign, toxin, carcinogen, teratogen (causing birth defects)—are simplistic and inaccurate. Moreover, they may result in policies that do more harm than good, as substances classified as unsafe in one context can improve health in another, or may replace substances causing even greater harm. For example, banning all carcinogens from the food supply—if it were possible—could cause more harm than good because fruits and vegetables contain significant quantities of naturally occurring compounds that, according to current understanding, are as carcinogenic as many synthetic pesticides.[4] Analogously, anticancer drugs with terrible side effects are tolerated when no more effective therapy is available.

An alternative approach relies on analysis to predict the consequences of different policies and compares them using society's values. Identifying improvements in environmental policy requires understanding of the physical and biological sciences to predict the effects of environmental pollution on human health and on ecosystems, of the social sciences to predict the effects of policy and environmental change on behavior, and of social values to identify preferences over a diverse set of desired attributes. Environmental pollution can cause a multiplicity of health problems, ranging from respiratory illness to neurological impairment and cancer. It can alter ecosystems, causing a change in the abundance and ecological roles of wildlife species, including both the introduction of exotic and the extinction of endemic species. It can alter aesthetic attributes, such as visibility. Concerns about social equity arise because environmental policies can alter the allocation of health, environmental quality, and financial costs across the citizenry. The natural sciences cannot resolve these "value" issues. They can be elucidated by the social sciences, but ultimately value choices must be resolved through the political system.

The complexity of environmental decisionmaking requires analysis to characterize the likely consequences of alternative options. Although decisions can be and often are made on the basis of holistic judgments or gut feelings, environmental issues are so complex that intuition can lead one astray. The management and decision sciences have developed analytic techniques to improve decisionmaking about complex matters—to make them more consistent with the decisionmakers' preferences. If the last generation of U.S. environmental policy was driven largely by a loud outcry against pollution, the next generation must be based on careful analysis of competing options, countervailing consequences, and public values.

Analysis to Support Environmental Decisions

Tools for analyzing environmental decisions have proliferated with growing recognition of the complexity of the issues. In addition to general policy-analytic methods like benefit-cost (often called cost benefit) and cost-effectiveness analysis, the new tools include risk assessment, comparative-risk analysis, risk-tradeoff analysis, risk-risk (or health-health) analysis, and life-cycle analysis. Similar in many respects, these species and subspecies of analysis focus attention on different aspects of a decision.

Risk assessment is a method for identifying the harms to human health or ecosystems that may result from a specified activity and then estimating the probability that such harms will occur and the number of people who may be affected. Although people have always assessed risks in their daily lives, the development of formal techniques was stimulated by interest in characterizing the risks of cancer associated with chemicals or radiation released to the environment and the risk of accidents associated with industrial facilities (for example, nuclear power plants).[5]

Risk assessment typically includes four steps: hazard identification (description of the harms that may result from exposure to the chemical or other agent); exposure assessment (determination of the number of people who will come in contact with the agent through air, water, soil, food, or other pathways and the quantities to which they will be exposed); dose-response assessment (estimation of the relationship between the quantity of the agent to which a person is exposed, or the dose a person ingests, and the probability of harm); and risk characterization (summary of the estimated probability and characteristics of possible harms, including relevant uncertainties and strength of evidence).

Standardized techniques for estimating the risks of carcinogenesis and industrial accident have been developed and adopted by the EPA and other regulatory agencies. Nevertheless, the accuracy and reliability of these techniques are uncertain and controversial. Uncertainty is inherent in the problems to which risk assessment is usually applied, such as predicting very rare events, with probabilities as low as one chance in a million that an individual will develop cancer sometime in his or her lifetime from exposure to a particular chemical. This probability, if nationally representative, would mean that about 250 of the people currently living in the U.S. would get cancer from this exposure. It would be virtually impossible, however, to identify these cases against a background of cancer arising from many other causes.

We cannot measure the probability of every risk in which we have an interest, in part because we do not consciously want to expose humans to them. Consequently, risk assessment for chemicals relies on exposing mice, rats, and other laboratory animals to much larger doses than humans face—doses where the risk to the animal is closer to one in two or one in ten—and extrapolating to the dose that would produce a very small risk to humans. Two extrapolations are required: the low-dose extrapolation, which requires assumptions about the shape of the curve relating dose to response (the probability of cancer), and the interspecies extrapolation, requiring assumptions about how large differences between rodents and humans in weight, life expectancy, metabolic rate, and other characteristics affect the dose-response relationship. Progress is being made in understanding and modeling the relationships between environmental exposures, resulting doses to target organs, and carcinogenesis; however, biochemistry and physiology continue to provide only limited guidance as to how these extrapolations should be made.[6]

In recent years, attention has been directed to adapting risk assessment to evaluate risks of other effects on human health (such as neurological and reproductive impairment) and effects on ecosystems. These efforts are less well developed and face challenges related to limited understanding of the physiological mechanisms of disease and, even more, of the interactions of species and resources in ecosystems.

Comparative-risk analysis (CRA) takes a broader perspective. It was developed as a method for identifying government priorities among environmental and other issues, and for allocating resources across issues. CRA typically provides an overall judgment of the relative importance of diverse environmental problems, such as air and water pollution, hazardous waste, and global climate change. Because such comparisons integrate quantitative information about the number of people or other organisms likely to suffer adverse effects with judgments about the relative severity of effects, CRA typically relies on informed judgment by expert or citizen panels. The first CRA studies were conducted by the EPA, and many states and localities have subsequently undertaken such studies.[7]

An important issue in CRA is the apparent conflict between expert and public assessments of the importance of dissimilar risks. On the one hand, experts often equate importance with the probability of harm and the number of people at risk. On the other hand, public perceptions seem to be influenced by different attributes, including familiarity with the source of the risk, whether it offers perceived benefits, and whether it is subject to control by the people at risk. The extent to which these differences in perspective reflect legitimate differences in values or errors of perception is subject to debate.[8]

Risk-tradeoff analysis (RTA) is a method for evaluating environmental decisions that attempts to highlight the risks that may be created by an activity intended to reduce risk. Echoing the economists' refrain that there is "no free lunch," risk analysts are increasingly discovering that actions intended to reduce one risk—the "target risk"—may create or exacerbate other, "countervailing" risks.[9] For example, cleaning up a contaminated waste site may require excavating tons of soil and transporting it to another site, which creates a (presumably smaller) risk of leakage and contamination at the new site. In addition, the excavation and transportation of soil create occupational risks to workers and traffic-accident risks to other road users, and may disperse some of the pollutants to the air. Similarly, disinfecting water with chlorine produces chloroform and other trihalomethanes that create a cancer risk for those drinking the water; failure to disinfect creates a risk of microbial contamination that can produce serious and potentially fatal gastrointestinal illness.

Risk-risk or *health-health analysis* highlights another pathway through which regulations intended to improve public health may actually worsen it, or at least lead to less improvement than expected. Aaron Wildavsky and others have articulated a relationship between economic resources and health, at both the family and social levels.[10] Drawing on this relationship, they argue that some share of the economic resources that are diverted from other uses by government programs—regulatory and other—would have been spent in ways that reduce health risk, such as buying new tires to reduce traffic risk. Assessments of reductions in health risks due to regulation need to incorporate this indirect increase in risk. Early empirical estimates suggest that the magnitude of government-imposed costs that induces one premature fatality is between $2 million and $70 million.[11]

Life-cycle analysis (LCA) is a method of comprehensively accounting for the environmental effects of a particular environmental decision, such as a choice among competing technologies. LCA attempts to include all environmental impacts of the technology, from acquisition and processing of raw materials through manufacturing, operation or consumption, to final disposal. LCA has been applied to evaluate the choice between alternative products—such as disposable paper and plastic coffee cups, shopping bags, disposable and cloth diapers, and also to evaluate the total impacts of electric automobiles, including the lead pollution associated with battery production and recycling.[12]

Cost-effectiveness analysis (CEA) and *benefit-cost analysis* (BCA) are economic methods that have comparatively long histories of application to health and environmental policy. CEA compares the cost of achieving similar goals by different methods; for example, the cost of different medical interventions to prolong survival after diagnosis of cancer. CEA does not provide a method for determining whether an action is beneficial overall, but only whether it is the least costly of the evaluated actions for achieving the selected end. CEA has been used to evaluate the effectiveness of alternative policies for reducing sulfate emissions from coal-fired electric plants, for example.[13]

In contrast, BCA attempts to account for all the consequences of available alternative actions—beneficial and adverse. In general terms, it is the inescapable framework for making difficult decisions, both public and personal.[14] In practice, of course, BCA faces certain limitations. Combining benefits and costs requires using some common metric; economists usually use money as the basis of translation. But there are substantial difficulties in reliably estimating economic values, which are the rates at which citizens are willing to trade more of one desired environmental or health attribute for less of another, or for less wealth to spend on other goods. Opponents of BCA argue that it is demeaning and even immoral to assign monetary values to health and the environment; proponents counter that such values are implicit in a decision, and decisions can be improved if the tradeoffs are acknowledged and debated openly.[15]

Methods for estimating the economic values of environmental attributes are an active research topic.[16] One tradeoff that has been the subject of much attention is the "value of a statistical life," defined as the willingness to pay to reduce a small risk of premature mortality (equivalently, the total that a large number of people would pay to eliminate a risk that is expected to randomly kill one among them). Currently recommended values approximate $3 million to $7 million.[17] These are based on estimates of the wage premiums paid to workers in hazardous jobs, consumer expenditures for smoke detectors and other safety devices, and surveys in which consumers are asked if they would purchase hypothetical risk reductions at specified prices; these estimates are of course sensitive to the assumption that workers and consumers accurately perceive the risk attributes of the choices they face.

BCA often does not take account of the distributional consequences of policies, which typically yield net benefits to some citizens and net losses to others, although the distribution of effects across the population can be described. One rationalization for not giving great weight to the distributional effects of a decision (if they do not violate fundamental rights) is that, if BCA is applied consistently, those who lose from one policy decision are likely to gain from another and, in the long run, everyone will be better off if policies are based largely on aggregated benefits and costs. Whether this rationalization is empirically supportable is open to doubt, however, and even if it were it might not justify the neglect of equity issues implicit in BCA. Another justification for not including distributional effects in BCA is that other social institutions, such as the tax and government benefits systems, may be better poised than the regulatory system to redress distributional inequities.

In summary, most advocates of BCA recognize that it should not be the sole criterion for policy decisions because it neglects distributional concerns and because methods for estimating benefits and costs are far from perfect; however, advocates argue that if estimated benefits and costs are explicitly considered, many bad decisions could be avoided, the public could more effectively understand and participate in regulatory policy debates, and better approaches to health and environmental protection could be identified and pursued.[18]

Whatever analytic framework is employed, the complexity of environmental issues and limitations in the science base ensure that the consequences of alternative policies can be predicted only with substantial uncertainty; often, predictions are uncertain to a factor of ten or more. Moreover, in many cases time-lags in detection and reversal of effects limit the possibilities for modifying policy decisions based on observed changes in the system. Both the large uncertainties and the limited potential for feedback to policy suggest that it is important to recognize and explicitly characterize uncertainties in predicted effects. A misplaced demand for (or assumption of) certainty in the predicted consequences of alternative policies can lead to poor decisions. Techniques for incorporating uncertainty in decision-oriented frameworks are moderately well developed and increasingly used in applications. These techniques include methods for estimating the value, in terms of improved decisions, of collecting additional data.[19]

Usable Knowledge: Science and Data to Support Analysis

These myriad types of analysis are frameworks for identifying the factors that should influence decisions, for organizing existing information, and for evaluating the types of additional information that could be collected. They are powerful tools for employing whatever information can be brought to bear on a decision. When little information is available, these methods can reveal that fact. To provide specific guidance in decisions, they require comprehensive and reliable knowledge about the environmental, health, and economic consequences of alternative policies and, for CRA, RTA, LCA, and BCA, the social values for evaluating tradeoffs among consequences.

The extent and quality of current data vary by topic. In general, the best quantitative information is available for pollutants released to the environment and their ambient concentrations in air and water. For example, time-series measurements of the ambient concentrations of ozone, particulates, and other "criteria" air pollutants designated by the Clean Air Act are of high quality and cover several decades for major urban areas. Data on human exposure to pollutants are less comprehensive; a small number of important studies have shown that exposure to most air pollutants is dominated by local, often indoor, sources, rather than by ambient outdoor levels.[20]

Data on the effects of exposure to humans and, especially, to ecosystems are most limited. The limited quality of data on human effects results in part because environmental-health effects are hard to distinguish from background variability (especially for effects like cancer that have long latency periods). For ecosystems, the difficulties involve the complexity of ecosystems, absence of a clear definition of "ecosystem health," and relative lack of attention and research funding for applications. Even seemingly elementary quantities, such as the number of species inhabiting the Earth, are poorly known. Recent estimates range from three million to thirty million.[21]

A series of comparable measurements taken over a long time is invaluable for basic natural science and epidemiological research directed toward understanding the relationships between pollution, environmental quality, and consequences, and for monitoring trends in order to evaluate the effects of policy. For example, various long-term measurements have been essential for investigating the contribution of carbon dioxide and other atmospheric gases to global climate change via the enhanced greenhouse effect. The availability of meteorological measurements covering large parts of the globe has allowed researchers to estimate the global annual mean surface temperature back to the 1860s and annual temperature for central England back to the seventeenth century. The estimated global series is compromised, however, by the paucity of data for regions with little population and ship traffic, such as the southern oceans. The continuous record of atmospheric carbon dioxide measured since 1957 from the top of Mauna Loa in Hawaii has provided incontrovertible evidence of the substantial increase in the concentration of this compound in the atmosphere. In contrast, the absence of comparable measurements for atmospheric aerosol concentrations has proven to be an important factor impeding progress in climate science.[22]

Similarly, although criteria air pollutants in major U.S. cities have been well measured for decades, advances in understanding the health effects of particulates suggest that small particles, which have not been distinguished in monitoring systems, are more relevant to health than the total particulates which have been measured.[23] Internationally comparable cancer and birth-defect registries are invaluable for analyzing environmental health threats;[24] measurements of lead in children's blood provided strong evidence for the beneficial effects of removing lead from gasoline.[25] The absence of geographically dispersed long-term data on human sperm counts means that there is little context for evaluating the implications of recent evidence of declines.[26]

More data alone are not always sufficient, and may be misleading, for improving environmental decisions. Summary measurements of pollutants that do not recognize differences in toxicity and possibilities for human exposure may encourage misdirected control efforts. Measurements of hazardous-waste production that do not distinguish the toxic components from the water with which they are mixed provide a poor indication of risk; large reductions in reported waste production may reflect only the concentration of hazardous components in smaller volumes of waste water. Toxic Release Inventory (TRI) data that aggregate pollutants released to the air with those transported to legal waste treatment and disposal facilities provide limited information about the risk to health and the environment. Extensions of the TRI concept to toxic-use inventories can further weaken the utility of the data for evaluating risk, since the connection between use of a toxic compound and risk is critically dependent on how the compound is used. For example, under the Massachusetts Toxic Use Reduction Act, a food producer was required to report (and consider options for reducing) its use of acetic acid, otherwise known as vinegar, which constitutes a primary ingredient in its salad dressing.[27]

Obstacles to Better Use of Analysis

The limited role analysis currently plays in environmental decisionmaking is neither accidental nor easily rectified. It reflects the structure of political and regulatory institutions and the preferences of interested parties. Because of significant uncertainty about the underlying science and often unavoidable tradeoffs between the interests of multiple stakeholders (including diverse groups of consumers, workers, managers, shareholders, and politicians), regulators are understandably reluctant to be explicit about the rationale for a decision. When someone's ox will be gored by any decision, it is tempting to hide behind the representation that the "science" compels the decision that has been made.[28]

Legal and institutional factors discourage a synoptic perspective and careful balancing of all the important consequences of alternative choices.[29] Most of the major enabling statutes are specific to a single environmental medium (air, surface water, soil, and groundwater) or pollutant stream (oil spills, pesticides, hazardous waste) and do not promote, or even allow, consideration of effects on other media or pollutant streams. Logically enough, the EPA is organized into program offices corresponding to the major statutes it is charged with enforcing. Decision rules and standards vary among the laws: the Clean Air Act requires that air quality standards be set "to protect the public health" while "allowing an adequate margin of safety"; the Federal

Insecticide, Fungicide, and Rodenticide Act provides that a pesticide should be approved for use if "it will perform its intended function without unreasonable adverse effects on the environment." Oversight authority is not centralized in a single congressional committee concerned with the environment as a whole, but dispersed to dozens of committees and subcommittees, each with its own parochial or tunnel vision. The diversity in objectives and oversight has led to apparently large and unjustifiable disparities in the marginal cost of health-risk reductions achievable in different venues, suggesting that greater health improvements could be achieved at the same cost by strengthening some standards and relaxing others.[30]

With a few exceptions (such as the Endangered Species Act and wetland provisions of the Clean Water Act), most of the major environmental statutes identify improvements in human health as a primary goal of environmental protection. Because of the political salience of human health, this linkage has undoubtedly led to stronger pollution standards than would have been imposed if the only goal were the preservation of environments and ecosystems in a "natural" state (even though environmental preservation might require zero pollution).[31] Moreover, it has arguably led to neglect of other environmental goals and obfuscation when proponents of environmental protection must justify their objectives by an appeal to human health.

Several attempts have been made to employ analysis to enhance consistency in decision making in the federal government. The National Environmental Policy Act requires the preparation of environmental impact statements that characterize the effects on the environment of proposed federal actions and alternatives. Since 1981, federal agencies have been required under Executive Orders 12291 (issued by President Reagan) and 12866 (issued by President Clinton) to conduct regulatory impact analyses of proposed regulations imposing annual costs of at least $100 million (or meeting other criteria of "major" impact). The Unfunded Mandates Reform Act (1995) requires executive agencies to conduct analyses of the benefits and costs of major rules and to justify selecting any but the "least costly, most cost-effective or least burdensome" option. It also requires the Congressional Budget Office to estimate the direct costs of certain legislation, thus beginning to impose the discipline of BCA and CEA on the legislative process.

Environmental advocacy groups have not in general supported increased use of analysis, despite achieving some notable successes when they have adopted it (for example, the EDF's evaluation of lead's impact on children's neurological development resulting in a phase-out of lead in gasoline). Reasons for their opposition vary, but often include an interest in directing attention to reducing pollution (for example, through pollution prevention) rather than balancing the benefits and costs of pollution control; a suspicion that current risk-analytic methods underestimate risks (because they underemphasize effects other than cancer and neglect possible synergy between pollutants); a concern that risk analysis is unethical, as it may encourage acceptance of the imposition of health risks on citizens (especially those who are disenfranchised by socioeconomic factors), and undemocratic, as it relies on specialized expertise; and a concern that environmental groups will be outgunned on an analytic battlefield by well-funded industry groups.[32] But, as already noted, actions to reduce one pollutant may cause greater harm than they prevent and, as discussed below, requiring documented analysis of the rationale for a decision opens the process to all interested parties and should impede efforts by special interests to exert undue influence.

It is widely believed that benefits of environmental regulation are more difficult to quantify than costs and that unquantified factors are underemphasized in a decision process that relies on analysis. Indeed, many proponents of an enhanced analytic role in decisionmaking have been attracted by the belief that it will lead to less regulation, either because of the difficulty of demonstrating that proposed regulations are beneficial or because imposing analytic requirements on regulatory agencies may prevent the issuance of new rules through "paralysis by analysis." Examination of past regulatory decisions shows that these hopes may be misplaced: analysis can lead to stronger as well as weaker regulation.[33]

Recommendations for the Next Generation

To reap the benefits a more analytic basis for environmental policy can provide, next-generation policymakers need to modify the institutions and incentives impeding thoughtful use of data, analysis, and science. One step would be to revise statutory decision criteria, agency structure, and congressional oversight mechanisms to promote acknowledgment and consideration of all the important consequences—environmental, health, and economic—of alternative policy choices. This might be achieved by piecewise reform of the existing legislation, by integrated enabling legislation, by a "supermandate" requiring analytic support for decisions or a "superauthorization"[34] permitting use of analysis despite underlying statutory prohibition. Although new enabling legislation or a supermandate is potentially more effective than piecewise reform, proposing such large-scale revision increases the risks to all factions that aspects of the current system they cherish will be lost; superauthorization is a more modest alternative.

Before he was appointed to the Supreme Court, Justice Stephen Breyer proposed the creation of an elite risk-analytic corps within the federal government.[35] This corps would provide a career path for civil servants to move among the risk-regulatory agencies (such as the EPA, Food and Drug Administration, Consumer Product Safety Commission, Occupational Safety and Health Administration), the Office of Management and Budget, and the congressional oversight committees. Members would be charged with the mission of promoting consistency in decisionmaking across contexts and would develop both substantive expertise and valuable personal relationships through their diverse professional assignments. However, the creation of such a group can be criticized as elitist and would not alter the balkanized statutory and legislative framework. Alternatively, a Council of Risk Advisors, like the Council of Economic Advisors, could be established with authority to support, review, and improve the use of risk analysis across the federal government, and a joint House-Senate Committee on Risk could be established to promote consistency in legislation concerning risk management.[36]

If analysis is to assume a larger role in decisionmaking, concerns about accessibility, credibility, and quality need to be addressed. In principle, a primary advantage of analytically based decisions is that the analysis documents the predicted consequences of the decision options considered and the rationale for the decision. Such documentation, if it is made public, allows interested parties to understand the justification for a decision; challenge and potentially correct errors of fact, omission of important factors, or implausible assumptions; and challenge the tradeoffs among effects that are either explicit or at least more easily identified than without such documentation. A documented analytic basis for decisions is consequently more democratic than a process relying on agency discretion, since it opens the decision process to public scrutiny and thereby limits the influence of groups with special access to regulators.[37]

Analyses should be subjected to serious review. Peer review by experts with some degree of independence from the decision should improve the analytic and scientific integrity of an analysis. The EPA's Science Advisory Board provides one model; alternatively, the Office of Science and Technology Policy could review risk assessments as the Office of Management and Budget reviews benefit-cost analyses.[38]

The analytic underpinnings of a decision should be (and are) subject to public review and comment under the Administrative Procedures Act. As in a legal case, proponents and opponents of a proposed decision can be expected to marshal evidence supporting and refuting debatable aspects of the analysis; indeed, parties whose interests are at stake have greater incentive to identify the weak points of an analysis than do "disinterested" experts. Analytic integrity is not the only basis on which advocates are likely to challenge a decision, of course; sound-bites and misrepresentation are powerful tools in any public debate. Nevertheless, holding the analytic high ground should rarely put an agency at a disadvantage, and the necessity to defend the analysis in public should stimulate improvements in methods for communicating the analysis and results to a diverse audience.

Reliance on documented analysis opens the decision process to competing analyses. Nongovernmental organizations (NGOs) representing all parts of the political spectrum could become providers of analysis. To some extent, competition among analyses already occurs; under the Superfund program, for example, firms that may be held liable for cleaning up contaminated sites sometimes conduct their own risk analyses as a basis for challenging the results of one by the EPA.

The existence of multiple analyses, conducted by or on behalf of groups with competing interests, may "illuminate the truth by triangulation."[39] Analytic competition should, over time, stimulate improvements in underlying data, analytic methods, and understanding by policy makers. Although environmental NGOs suggest they cannot compete with well-funded industry groups, credibility is not necessarily enhanced by superiority of economic resources. Resource limitations might be mitigated by modifying provisions in existing legislation that allow NGOs to be awarded litigation costs to support development of competing analyses.

Decision rules should be modified to provide incentives for interested parties to develop improved information. In its risk assessment of potentially carcinogenic substances, the EPA has been criticized for being unwilling to depart from "default assumptions" that should be made only when context-specific data are unavailable. If agencies are unwilling or unable to incorporate new data in their decisionmaking, outsiders will have little incentive to generate them.

Shifting the burden of proof or the default decision option can create powerful incentives for improved decisionmaking. Often, an industry facing more stringent regulation can benefit by delaying a decision, typically by arguing that "there is too much uncertainty" about the scientific issues to support new regulation. Shifting the default outcome can provide the industry with an incentive to accelerate rather than delay a regulatory decision. When a new product cannot be introduced without regulatory approval, as in the case of pharmaceuticals, medical devices, and some newly synthesized chemicals under the Toxic Substances Control Act, manufacturers do not argue that there is too much uncertainty about the effects of their newly developed products; rather, they generate and provide relevant data.

A potentially useful model for regulating existing products and activities is the California Safe Drinking Water and Toxic Enforcement Act of 1986 (Proposition 65). This law requires that firms exposing workers or members of the public to any compound identified by a state advisory board as causing cancer or reproductive toxicity must either provide a warning or demonstrate that the risk is less than an administratively determined level which presents "no significant risk." There has been little systematic evaluation of the law, but it appears to have hastened the development of regulatory standards for many chemicals in California and led to removal of several consumer-product risks without creating significant economic burden.[40] Although the decision standard ("no significant risk") does not allow balancing benefits and costs, the possibility that the law will impose excessive costs is limited by a firm's option to provide a warning if it cannot show its product to be "safe" (a few products, such as paint removers, carry warnings). Alternatively, a law could be crafted where the default option is to prohibit environmental release unless the regulatory agency (or producer) can show that the costs of preventing release are likely to exceed the benefits. Such a law would combine incentives for industry to develop relevant information with a decision standard that requires comparing benefits and costs.

Over the long term, the continued development of environmental science and the establishment and maintenance of long-term environmental data series are crucial to improved decisionmaking. The EPA's support for science is viewed as too responsive to immediate concerns—the "risk-of-the-month" syndrome—to sustain the development of a strong science base, although administrative changes have been made within the agency to provide more stable support. Emerging issues not in the domain of existing regulatory offices may receive inadequate attention. One proposal to improve the development and application of science is to create the position of chief scientist within the EPA, with authority over research funding and the responsibility to improve the quality and application of science within the agency. Alternatively, separation of the research and regulatory missions may be useful, for example, by creating a National Institute for the Environment. A national institute could support a more stable and coherent research program, promote integration across environmental media and regulatory areas, and support research in areas falling between regulatory program areas.

Designing an appropriate organization to direct research is challenging. Too much control by short-term regulatory agendas limits the consistent support needed for long-term development, but providing too much latitude for curiosity-based research may yield a program that is expensive and slow to address matters of regulatory importance. The goal is to design an organization that can maintain an unstable equilibrium between the needs of short-term regulatory agendas and the long-term requirement for broad and deep understanding of issues central to environmental decisionmaking.[41] An alternative may be to divide environmental research, entrusting regulatory agencies with responsibility for efforts targeted to current needs and the National Science Foundation, National Institutes of Health, or a National Institute for the Environment with responsibility for topics where the benefits are more distant.

Whatever the institutional setting, research priorities and analytic methods can be improved by retrospective analysis of environmental decisions and ongoing monitoring and assessment of their effects. The substantial uncertainties inherent in predicting consequences of environmental decisions warrant humility and an eagerness to learn from experience.

Environmental regulation influences a substantial share of the U.S. economy. It is too important, and the issues are too complex, to rely on intuition and unaided judgment to determine what changes will most efficiently improve environmental quality, human health, and national welfare. The next genera-

tion of environmental policy needs to build on existing methods and practices of analysis to support decisionmaking, to stimulate the generation of
usable knowledge, and to enhance the role of analysis in decisionmaking.
Doing so will require changes in the legal and institutional framework in
which environmental regulation occurs, replacing the current balkanized
laws, regulatory program offices, and congressional oversight committees
with structures that support a more integrated, synoptic perspective and promote careful balancing of diverse and often competing objectives.

Notes

1. EPA, *Environmental Investments: The Cost of a Clean Environment* (Washington,
D.C.: Government Printing Office, 1990).

2. Environmental Risk Reduction Act of 1993, S. 550, 103d Congress.

3. Quoted in Joseph V. Rodricks, *Calculated Risks: The Toxicity and Human Health
Risks of Chemicals in Our Environment* (Cambridge: Cambridge University Press, 1992),
39.

4. National Research Council, *Carcinogens and Anticarcinogens in the Human Diet*
(Washington, D.C.: National Academy Press, 1996); B. N. Ames, "Dietary Carcinogens
and Anticarcinogens," *Science* 221 (1983): 1256–64.

5. Rodricks, *Calculated Risks;* Vincent T. Covello and Miley W. Merkhofer, *Risk
Assessment Methods: Approaches for Assessing Health and Environmental Risks* (New
York: Plenum Press, 1993); C. D. Holland and R. L. Sielken, Jr., *Quantitative Cancer
Modeling and Risk Assessment* (Englewood Cliffs, N.J.: Prentice-Hall, 1993); National
Research Council, *Risk Assessment in the Federal Government: Managing the Process*
(Washington, D.C.: National Academy Press, 1983); National Research Council, *Science
and Judgment in Risk Assessment* (Washington, D.C.: National Academy Press, 1994).

6. S. Olin et al., eds., *Low-Dose Extrapolation of Cancer Risks: Issues and Perspectives*
(Washington, D.C.: International Life Sciences Institute Press, 1995).

7. See EPA, *Unfinished Business: A Comparative Assessment of Environmental Prob-
lems* (Washington, D.C.: Government Printing Office, 1987); EPA, *Reducing Risk: Setting
Priorities and Strategies for Environmental Protection* (Washington, D.C.: Government
Printing Office, 1990); Finkel and Golding, eds., *Worst Things First?* (Washington, D.C.:
Resources for the Future, 1994); and J. C. Davies, ed., *Comparing Environmental Risks:
Tools for Setting Government Priorities* (Washington, D.C.: Resources for the Future,
1996).

8. B. Fischhoff et al., "How Safe Is Safe Enough? A Psychometric Study of Attitudes
Toward Technological Risks and Benefits," *Policy Sciences* 8 (1978): 127–52; P. Slovic,
"Perceptions of Risk," *Science* 236 (1987): 280–85; H. Margolis, *Dealing with Risk: Why
the Public and the Experts Disagree on Environmental Issues* (Chicago: University of

Chicago Press, 1996); C. R. Sunstein, "Which Risks First?" *University of Chicago Legal Forum* (forthcoming).

9. J. D. Graham and J. B. Wiener, eds., *Risk versus Risk: Tradeoffs in Protecting Health and the Environment* (Cambridge, Mass.: Harvard University Press, 1995).

10. A. Wildavsky, "No Risk Is the Highest Risk of All," *American Scientist* 67 (1979): 32–37; A. Wildavsky, "Richer Is Safer," *Public Interest* 60 (1980): 23–39.

11. W. K. Viscusi, "Risk-Risk Analysis," *Journal of Risk and Uncertainty* 8 (1994): 5–17; W. K. Viscusi, "Mortality Effects of Regulatory Costs and Policy Evaluation Criteria," *Rand Journal of Economics* 25 (1994): 94–109. See also R. L. Keeney, "Mortality Risks Induced by Economic Expenditures," *Risk Analysis* 10 (1990): 147–59; R. L. Keeney, "Estimating Fatalities Induced by the Economic Costs of Regulations," *Journal of Risk and Uncertainty* 14 (1997): 5–23; and W. K. Viscusi, ed., "Risk-Risk Analysis Symposium," *Journal of Risk and Uncertainty* 8 (1994 [special issue]): 5–122.

12. M. A. Curran, ed., *Environmental Life-Cycle Assessment* (New York: McGraw-Hill, 1996); P. R. Portney, "The Price Is Right: Making Use of Life Cycle Analyses," *Issues in Science and Technology* 10, no. 2 (1993): 69–75. See also chap. 2 above.

13. M. R. Gold et al., *Cost-Effectiveness in Health and Medicine* (New York: Oxford University Press, 1996).

14. E. Stokey and R. Zeckhauser, *A Primer for Policy Analysis* (New York: W. W. Norton, 1978); T. Tietenberg, *Environmental Economics and Policy* (New York: Harper-Collins, 1994); N. Hanley and C. L. Spash, *Cost-Benefit Analysis and the Environment* (Brookfield, Vt.: Edward Elgar, 1993); R. E. Just, D. L. Hueth, and A. Schmitz, *Applied Welfare Economics and Public Policy* (Englewood Cliffs, N.J.: Prentice-Hall, 1982).

15. C. R. Sunstein, "Incommensurability and Valuation in Law," *Michigan Law Review* 92 (1994): 779–861.

16. A. M. Freeman III, *The Measurement of Environmental and Resource Values: Theory and Methods* (Washington, D.C.: Resources for the Future, 1993).

17. W. K. Viscusi, *Fatal Tradeoffs: Public and Private Responsibilities for Risk* (New York: Oxford University Press, 1992).

18. K. J. Arrow et al., "Is There a Role for Benefit-Cost Analysis in Environmental, Health, and Safety Regulation?" *Science* 272 (1996): 221–22; Robert Hahn, ed., *Risks, Costs, and Lives Saved* (New York: Oxford University Press, 1996).

19. H. Raiffa, *Decision Analysis: Introductory Lectures on Choices Under Uncertainty* (Reading, Mass.: Addison-Wesley, 1968); M. G. Morgan and M. Henrion, *Uncertainty: A Guide to Dealing with Uncertainty in Quantitative Risk and Policy Analysis* (Cambridge: Cambridge University Press, 1990); R. M. Cooke, *Experts in Uncertainty: Opinion and Subjective Probability in Science* (New York: Oxford University Press, 1991); J. K. Hammitt and J. A. K. Cave, *Research Planning for Food Safety: A Value-of-Information Approach* (Santa Monica, Calif.: Rand, 1991); P. R. Kleindorfer, H. C. Kunreuther, and P. J. H. Schoemaker, *Decision Sciences: An Integrative Perspective* (Cambridge: Cambridge University Press, 1993); K. M. Thompson and J. D. Graham, "Going Beyond the Single Number: Using Probabilistic Risk Assessment to Improve Risk Management," *Human*

and Ecological Risk Assessment 2 (1996): 1008–34.

20. L. A. Wallace, *The Total Exposure Assessment Methodology (TEAM) Study: Summary and Analysis* (Washington, D.C.: EPA, 1987).

21. R. M. May, "How Many Species Inhabit the Earth?" *Scientific American* 267, no. 4 (1992): 42–49.

22. J. T. Houghton et al., eds., *Climate Change 1995: The Science of Climate Change* (Cambridge: Cambridge University Press, 1996).

23. R. Wilson and J. Spengler, eds., *Particles in Our Air: Concentrations and Health Effects* (Cambridge: Harvard University Press, 1996).

24. National Research Council, *Environmental Epidemiology* (Washington, D.C.: National Academy Press, 1991).

25. U.S. Department of Health and Human Services, *Second National Health and Nutrition Examination Survey (NHANES II)* (Washington, D.C.: Government Printing Office, 1976–1980).

26. T. Colborn, J. P. Myers, and D. Dumanoski, *Our Stolen Future: How We Are Threatening Our Fertility, Intelligence and Survival: A Scientific Detective Story* (New York: Dutton/Signet, 1996).

27. George Gray, vice chair, Massachusetts Toxic Use Reduction Act Science Advisory Board, personal communication.

28. W. E. Wagner, "The Science Charade in Toxic Risk Regulation," *Columbia Law Review* 95 (1995): 1613–1723.

29. J. B. Wiener and J. D. Graham, "Resolving Risk Tradeoffs," in Graham and Wiener, *Risk versus Risk.*

30. Stephen Breyer, *Breaking the Vicious Circle: Toward Effective Risk Regulation* (Cambridge: Harvard University Press, 1993); T. O. Tengs and J. D. Graham, "The Opportunity Costs of Haphazard Social Investments in Life-Saving" and Robert Hahn, "Regulatory Reform: What Do the Government's Numbers Tell Us?" in Hahn, *Risks, Costs, and Lives Saved.*

31. J. B. Wiener, "Law and the New Ecology: Evolution, Categories, and Consequences," *Ecology Law Quarterly* 22 (1995): 325–57.

32. Alon Tal, "A Failure to Engage," *Environmental Forum* 14, no. 1 (1997): 13–21; Daniel C. Esty, "What's the Risk in Risk?" *Yale Journal on Regulation* 13 (1996): 603–12.

33. J. D. Graham and J. K. Hartwell, eds., *The Greening of Industry: A Risk Management Approach* (Cambridge: Harvard University Press, 1997); R. D. Morgenstern, ed., *Economic Analyses at EPA: Assessing Regulatory Impacts* (Washington, D.C.: Resources for the Future, forthcoming).

34. J. B. Wiener, testimony before Senate Committee on Government Affairs, 8 March 1995; C. R. Sunstein, "Congress, Constitutional Moments, and the Cost-Benefit State," *Stanford Law Review* 48 (1996): 247–309.

35. Breyer, *Breaking the Vicious Circle.*

36. Wiener and Graham, "Resolving Risk Tradeoffs."

37. F. B. Cross, "Why Shouldn't We Regulate the Worst Things First?" *New York*

University Environmental Law Journal 4 (1996): 312.

38. Harvard Group on Risk Management Reform, "Reform of Risk Regulation: Achieving More Protection at Less Cost," *Human and Ecological Risk Assessment* 1 (1995): 183–206.

39. Prof. Daniel C. Esty, presentation to Harvard University Risk and Decision Sciences Seminar, 21 Feb. 1997.

40. C. M. Shulock, *Summary Report on Panel Meetings, Proposition 65 Review Panel* (Sacramento: California Environmental Protection Agency, 20 Feb. 1992); K. W. Kizer, T. E. Warriner, and S. A. Book, "Sound Science in the Implementation of Public Policy: A Case Report on California's Proposition 65," *Journal of the American Medical Association* 260 (1988): 951–55; W. S. Pease, "Chemical Hazards and the Public's Right to Know: How Effective Is California's Proposition 65?" *Environment* 33 (1991): 10–22.

41. Prof. M. Granger Morgan, head, Department of Engineering and Public Policy, Carnegie Mellon University, personal communication.

Toward Ecological Law and Policy

E. Donald Elliott

Most of today's environmental law violates the basic principles of ecology. Nature teaches the connectedness of all activities, but most current-generation law regulates separate pollutants with little consideration of ecosystems as a whole. The continuums of nature generally adapt gradually, but today's environmental law makes sharp distinctions between safe and unsafe, attainment versus nonattainment areas, permissible versus impermissible levels of pollution.

Instead of shaping industrial adaptation with incentives, today's federal pollution-control laws usually set federal "standards" as absolute legal edicts that brook no local exceptions. Rather than empowering decentralized authorities to adapt standards to local conditions, today's environmental law is premised too often on the fiction of an omniscient center—and depends too heavily on control at the federal level as a precondition to action. Many of the problems this structure has created are identified in chapter 1. This chapter will explore how the legal system in particular can be moved in a more ecological direction. It will also review more flexible and cost-effective ways to achieve environmental protection.

The claim that most U.S. environmental law is not ecological in its structure and philosophy does not apply equally to all aspects of today's legal policies. For example, the National Environmental Policy Act (NEPA), which mandates assessments of the effects on the environment of major federal actions, cannot be said to be anti-ecological in philosophy, only cumbersome and ineffective in implementation. Nor are those parts of national policy that strive to preserve portions of the environment as wilderness for future generations contrary to the ecological worldview in the same way as the great pollution control statutes of the 1970s and 1980s. The pollution-control statutes that sought to "clean up" the air, water, and land by setting legal standards to regulate pollution at the federal level are the centerpiece of current policy both in terms of cost and controversy.

Ordinances to control smoke, or to manage wastes, have existed in big cities like London and Chicago for hundreds of years. Although science increased public awareness of environmental pollution and the urgency for dealing with it in the 1960s, what is really distinctive about the "environmental law" created in the United States in the 1970s and 1980s are the legal techniques used for combating pollution. Largely the brainchild of the late Sen. Edmund Muskie of Maine, who was the principal author of several of the major federal environmental laws, including the Clean Air Act Amendments of 1970, the existing legal system for regulating pollution in the United States consists of the following elements:

- pollutant-by-pollutant, or industry-by-industry regulation carried out by government under federal statutes after lengthy administrative proceedings and court challenges;
- minimum standards set by administrative agencies at the federal level limiting the amount of pollution that may be put into the air, water, or land;
- requirements that the states translate federal goals into facility-specific legal requirements for individual factories or other sources of pollution;
- establishment of legal rights for environmentalists and other citizen groups to sue to enforce pollution laws, including bringing actions to force the government to act by deadlines specified in the law.

This basic system of legally mandated pollution reduction programs is often called the command-and-control system, because government both commands the degree of pollution reduction and controls (at least to some extent) the means for achieving these targets.

The federal pollution-control effort in the United States represents a major investment of social resources, in the range of $100 to $150 billion per year[1]—or approximately the same percentage of GNP as the Marshall Plan to rebuild Europe following World War II. It stands as a fundamental social commitment to restructure the economy to be more compatible with nature. Ironically, however, it is built on the nineteenth-century model of a machine or an early twentieth-century corporation—with the center issuing commands to the regional nodes for implementation. This system, called "cooperative federalism" during its genesis in the Nixon Administration and derided today in some quarters as "unfunded federal mandates," has, despite its flaws, been remarkably successful in achieving measurable progress in cleaning up the environment. But despite the success of the first generation of modern environmental law in controlling the growth of pollution in the areas that are currently regulated, there is a broad, bipartisan consensus that this model is not adequate for the next generation of environmental problems.[2] The command-and-control model is difficult to adapt to the diversity of problems and circumstances—and the concomitant need for a diverse set of tools and governmental activities at a variety of levels—that we face in the years ahead.

There are several reasons for the strong and growing consensus that we must reinvent and rejuvenate the legal tools used for environmental protection. First, many of the remaining environmental problems we face are not easily amenable to command-and-control approaches. Command and control by government bureaucracy has been compared (by Prof. Richard Stewart) to central planning of the economy.[3] It was developed to regulate a few large targets, industrial "polluters"—the so-called "big dirties," like power plants, refineries, chemical plants, and the automobile industry—which have dominated environmental law in its first generation. But as many of the other chapters in this book demonstrate, future environmental problems are often centered in other sectors of the economy, such as transportation, agriculture, and services, and in other practices, such as consumer lifestyles and consumption patterns. These are virtually untouched by the existing command-and-control policies and have proved very difficult to influence using command-and-control techniques. Among other problems, there is often a strong political backlash when government orders or prohibitions are perceived as restricting consumer choices or lifestyles—as demonstrated by several unsuccessful attempts to reduce automobile pollution not only by redesigning cars but also by reducing vehicle miles traveled. As we move beyond just regulating large utilities, factories, and other centralized sources of pollution to address problems caused by smaller and more diffuse sources of pollution, it becomes more difficult to utilize the traditional regulatory approach.

In addition, many believe that the current mechanisms for regulating pollution are unnecessarily blunt or cumbersome, resulting in economic costs that are much higher than they need to be. The high costs of government action also result in leaving a very large number of pollution sources "outside the system," so we simultaneously suffer from problems of overregulation of certain sources and little or no regulation of others. The present system also tends to shortchange science in favor of politics in focusing public attention on environmental problems. The present political system of environmental regulation is highly sensitive to certain voices (well-organized, technically sophisticated pressure groups, including environmentalists, as well as business people), but other voices without a lobby are largely left out of the dialogue (disadvantaged economic and racial minorities, disinterested scientists).

Strategies for Reform

There are serious divisions and disagreements about what direction legal policies and institutions for environmental regulation should take over the long run. On the one hand, some argue that the basic premises for federal regulation of the environment have been a mistake, and that we should return to a market-based system of environmental protection, perhaps through an improved system of property rights. This view, represented by some elements of the Republican Congress that came to power in the 1994 midterm elections, has not to date commanded a majority of the American people. The desire for more responsive, more efficient, and less expensive environmental law led by government is real, but the public wants better and smarter environmental regulation, not a return to the unregulated markets of the nineteenth century.

At the other extreme from the market theorists are visionaries who would redraft our environmental laws from scratch in a new "unified" or "organic" statute. Many interesting ideas have been suggested to form the basis for a new vision of environmental law and policy, some of which are summarized below along with the criticisms they provide. This chapter (like chapter 1) will ultimately adopt the perspective that we are not yet ready to commit ourselves to any one of these sweeping reforms, but that we must adopt an "evolutionary strategy" to environmental reform in which we continue to experiment with all of the innovative ideas—just as nature rarely commits itself to a single strategy, but maintains diversity and experimentation. Like nature, we can feel our way by trying a diversity of techniques on a small scale and adapting as we learn which are most successful. The ultimate problem, then, to which I will return at the end of this chapter, is how to maintain what is good and successful in what we have done to date in environmental law, while adapting the system to capture the potential to do better through the innovativeness and experimentation of the suggestions for reform.

Before turning to the immediate problem of what we should do, and how we transition to the next generation of environmental law, it will be helpful to review briefly a multiplicity of visions of where we should be going in the long run in the form of various competing ideas for the fundamental reform of environmental law.

A Menu of Environmental Reforms

Some of the promising innovative techniques for reforming environmental laws fall into four broad categories: economic incentives, improving environmental information programs, private or voluntary programs, and structural changes to environmental programs.

Economic Incentives

Marketable rights trading systems. The acid rain trading program instituted by the Clean Air Act Amendments of 1990 has provided a successful, large-scale demonstration that systems of tradable rights can achieve equivalent (or better) pollution control at a fraction of the cost of conventional systems under certain conditions (primarily the technical feasibility of accurately monitoring the pollution released). Tradable rights systems have long been advocated by economists as a more efficient way to control pollution, but opposed by some environmentalists as "licenses to pollute," which are morally unacceptable at least where issues of health are involved (see chapter 7). A "hybrid" or two-tiered system of regulation, with firm command-and-control limits to protect health, coupled with incentive systems, such as marketable rights, to create incentives for further reductions, offers a partial solution to this problem.[4]

Materials taxes. Rather than continuing the present policies of subsidizing virgin materials, and then complaining that recycled sources are "too expensive," some advocate using the tax system to create broad compensating incentives for raw materials policies that are less wasteful and more protective of the environment. Despite the strong intellectual appeal of these arguments, they have faced tough political opposition that so far has prevented their adoption in the United States. However, increasing understanding of the multiple benefits of taxing the use of goods and services rather than income, coupled with growing international pressure for common action, may eventually lead to some change in United States policies in this area.

Emissions taxes. Another broad incentive to reduce emissions into the environment would be emissions taxes, which could be based on existing indexes of pollution, such as the toxic release inventory (TRI) or other indicators tailored for that purpose. Despite significant environmental logic, there is again strong political opposition. A private alternative is developing as large purchasers, including the federal government, request that suppliers provide information about their practices, including use and release of chemicals, and then consider this information in making purchasing decisions.

Subsidies. Another economic incentive system that can sometimes be effective is government subsidies to promote pollution control, such as the public funding of sewage treatment plants (POTWS). Again, political considerations militate against broad-scale use of these techniques, but some narrow applications are increasing, such as using government and private procurement preferences to encourage the development of products and services with lesser adverse effects on the environment.

Loan programs. A variation of the subsidy strategy is to build environmental considerations into the qualification process for government programs. This is currently done for several programs, such as financing of large development projects through the government-owned Overseas Private Investment Corporation (OPIC) and the Export-Import Bank. Such an approach, designed to induce adherence to certain environmental standards, might be expanded to other programs, both at home and abroad.

Partial compensation. Another variation of subsidies is to provide partial compensation or incentives to owners of environmentally sensitive properties (such as wetlands) that are regulated for the general benefit. This concept is highly controversial when (allegedly) required as a matter of constitutional law under the takings clause of the Fifth Amendment. However, partial compensation as a matter of grace to those who are asked to make disproportionate sacrifices for the good of the community would reduce opposition to strong environmental programs (see chapter 3).

Improving Environmental Information Programs

"Information central." As computerized data about the environment increase geometrically, there is a great need to make data more meaningful and usable through standardizing and integrating environmental data systems. One proposed step in this direction is an independent federal Bureau of Environmental Statistics modeled on the Bureau of Labor Statistics. However, the concept of data integration is even broader and encompasses integrating private, facility-specific environmental data, developing more meaningful indicators of environmental performance, and making this data available to consumers and others for use in their decisions. The challenge of data integration also has international dimensions, as duplicative (and, potentially, unduly trade-restrictive) product testing and standards will gradually have to give way to more harmonious integrated systems of environmental testing for products that move across national boundaries, and to integrated systems for measuring and limiting pollution of global commons.

Self- or third-party certification and standard setting. Under the traditional command-and-control system, government processing of information can become a significant bottleneck that stifles effectiveness, both in setting environmental standards and in enforcing them. Increasingly, the processing and evaluating of environmental information will take place in decentralized nodes, with the role of government becoming setting the standards and spot-checking the system, rather than processing information at the "retail level." Thus, rather than the federal government attempting to regulate chemicals by analyzing all available scientific information on a chemical-by-chemical basis, the government could regulate at the "wholesale" level by setting protocols and criteria for private consensus standards (such as those set by ISO, ASTM), perhaps even establishing when such standards will be recognized as authoritative. In addition, increased use of standards set privately such as the ISO

14000 environmental requirements will provide benchmarks for performance. And eco-labels may give consumers the ability to choose environmentally preferable products.

A similar evolution is taking place in enforcement, as experiments with self-certification or third-party certification of environmental results advance. This approach to enforcement resembles what the Securities and Exchange Commission does to monitor corporate finances, eliminating the need to closely supervise every company and allowing limited government oversight resources to be focused on problem cases. Similar "reforms" have been proposed for the Food and Drug Administration area, but proved controversial.

Helpful information. Much of the past generation of environmental programs has been predicated on a threatening or punitive relationship between government and the regulated community. Experience has shown, however, that providing helpful information to the regulated community about the extent and source of harms and how to reduce environmental impacts can be a highly effective supplement to traditional regulatory programs. These "positive" programs are especially effective when parties already have strong incentives to address environmental issues for other reasons. Examples include the EPA's radon and indoor air quality programs, and OSHA's "Star" program, as well as traditional government programs in other areas, such as agricultural extension services.

Awards. Awards that provide positive recognition for extraordinary environmental innovations and performance can provide useful positive incentives. However, to reduce fraudulent or spurious claims (so-called Greenwash), government and/or third parties must be available to verify claims. The FTC's Guidelines on Environmental Marketing Information provide a good starting point.

Private Programs

Increasingly, the distinction between "private" and "public" programs is breaking down in the environmental area, as both norm setting and implementation of environmental programs are fueled by a host of important quasi-public initiatives.

Product stewardship. Product manufacturers are increasingly using life-cycle analysis to evaluate the environmental effects of their products after they are sold, and designing products to be easily recyclable or to incorporate reusable or "source-reduced" components that produce less waste or harm to the environment. Some countries are promoting the legal concept that manufacturers are responsible for the environmental fate of their products, and must implement "take-back" or other recycling schemes. Whether or not this is recognized as a formal legal principle in the United States, environmentally conscious consumers and organizations, as well as liability concerns, are increasingly leading manufacturers to factor the "downstream" effects of their products into product design decisions.

Environmental TQM. Many organizations are using advanced management and system design techniques such as total quality management to involve employees throughout the organization in redesigning products and production systems to prevent pollution and enhance environmental performance. The future of environmental law and regulation should be designed to facilitate and enhance the trends toward improved environmental management.

Structural Changes to Environmental Programs

Unified, cross-media statute. At present, the statutory authority for various federal environmental programs is spread out in dozens of separate statutes with differing language and substantive provisions. This not only promotes confusion and complexity but leads to "single-medium" thinking, so that environmental programs are sometimes designed to minimize effects on the air or water without adequate consideration of effects on other parts of the environment. The EPA has been moving for some time to regulate more on a "multimedia" basis. A movement has been developing that would go beyond current multimedia approaches to draft a single federal environmental "organic act" that would unify environmental laws into a single and consistent statutory structure. Opinions differ as to how much difference the revision of environmental laws would make in practice, as well as to the political practicality of recodifying existing laws.

Simplification. Today's environmental laws and regulations contain literally millions of words, and millions of dollars are spent every year interpreting them. Some reformers believe that a radical simplification of environmental requirements would improve the overall performance of the system.

An Evolutionary Strategy for Reforming Environmental Law

However appealing some of these ideas may be in the abstract to their proponents, it is unlikely that we will suddenly decide to sweep away the existing system of cumbersome and expensive—but effective—federal regulatory requirements developed during the 1970s and 1980s. A realistic reform strategy must retain and build on the successes of the past, and use them in evolving toward a next generation of legal technologies for environmental regulation. Nature shows us the way in the process of "epigenesis," by which an existing system transforms itself by performing new functions. Just as biological structures can evolve to perform new functions, it is possible to preserve the accomplishments of the first generation of environmental law, while simultaneously transforming the system into one that is wiser, more efficient, and more adaptable.

The Evolution of Environmental "Bubbles"

One way to transform the environmental law system is to extend gradually the existing "bubble" concept to allow broader, multimedia trading across various types of environmental risks. The bubble is a simple concept. Bubbles allow flexible, decentralized implementation of national mandates; they are like natural adaptation in that by enabling trading of environmental benefits across broader categories, they permit adaptation to local conditions or lower-cost opportunities.

Originally, the EPA developed the bubble concept by changing its definition of the "source" of air pollution that was regulated under the Clean Air Act from a single piece of machinery or emission point to encompass an entire factory from fence line to fence line. This broader definition of source had the effect of putting the entire factory under a single, imaginary "bubble" and thereby allowing "trading" between one source of emissions and another, as long as the total amount of pollution coming from the factory as a whole did not increase.[5] By controlling pollution more from one process, and less from another, a factory could achieve the same total level of pollution control at far lower costs. The original bubble policy was, however, subject to a number of restrictions that limited its usefulness: it applied only to a single type of pollution, and generally only to contiguous sources under common management control, like a single factory (although later some "banking" and trading with other sources was permitted under limited conditions).

The next major step in the expansion of the "bubble" logic was the acid rain trading program created by the 1990 Amendments to the Clean Air Act (see chapter 14). This statutory provision imposed roughly a 50 percent reduction (ten million tons) in pollution of sulfur dioxide over ten years from large power plants nationwide. By allowing trading among sources, the reduction is being accomplished at a fraction of what it would have cost to make the same reduction using traditional regulatory techniques. The "cap and trade" approach of the acid rain trading program essentially put a bubble over all the large power plants in the United States, and extended it over ten years. This was a significant expansion of the bubble concept in both time and space, but it was still limited to a particular kind of pollution coming from a limited category of sources.

A logical extension of bubbles would be to permit broader trading across different types of pollution. The expansion of multimedia bubbles (that is, trading of pollution control obligations across different kinds of pollution) brings with it the strong promise of greater environmental benefits at lower costs. The bubble concept furthermore allows us to draw into our environmental programs those who currently fall outside but whose activities may be the source of considerable harm. Point sources can contract with nonpoint sources who may be able to reduce their emissions at lower cost. Manufacturing companies could trade with services companies. One can even imagine, borrowing a concept from industrial ecology, drawing a bubble from a life-cycle perspective over suppliers and producers, or even producers and consumers. Once the government has established the risk exposure end points, those best positioned to reduce pollution should be given an incentive to act. There will, of course, be some significant challenges in implementation. It has long been recognized that trading one kind of pollution for another poses a host of practical problems, such as how to measure reductions in one type of pollution on a common basis with reductions in some other type of pollution. These problems, although difficult in theory, can be solved in practice in some clear cases by maintaining the present system of government-set command-

and-control standards as a benchmark, but allowing individual pollution sources to enter into enforceable agreements for substitute compliance where (i) the alternative pollution controls are clearly better for the public[6] and (ii) the source bears the burden of proving and monitoring (including the government's oversight costs) to demonstrate that the environment is better off.

Some believe that the difficulty of measuring various environmental risks counsels against a multimedia bubble that would permit far-reaching risk trade-offs. But this argument confuses measurement and comparison (or, in more technical terms, confuses cardinal and ordinal comparisons): even if one cannot measure something exactly, it may still be possible to make gross comparisons where the differences are large enough. We may not know exactly how small Vermont is, but we still know that California is larger. Similarly, although we cannot measure exactly, and therefore cannot make trade-offs between many environmental risks or benefits in close cases, this is not a problem in other cases if the differences are large enough. Some tradeoffs of one kind of pollution effect for another clearly represent a good investment from society's standpoint. By allowing trade-offs between various types of pollution risks, but only when it can be demonstrated clearly that one is greater than the other, we can also create incentives for the development of better techniques for comparison over time.

At a theoretical level, the logic of a bubble, or cap-and-trade, system is that which led the Nobel Prize–winning economist Ronald Coase to observe that the ability to "contract around" inefficient government rules was limited only by the costs of transactions.[7] Ultimately, therefore, if parties are free to trade, they will tailor the regulatory burdens to the local situation by transferring the burdens to where they can be borne most cheaply—and thereby we can attain much greater protection of the environment for the same investment of resources.

The advantage of the multimedia bubble approach is not only lower costs but the creation of incentives to control types and sources of pollution or environmental harm that are currently outside the ambit of government regulation. These presently unregulated sources may be difficult to regulate via command-and-control approaches, but can be induced to make changes to benefit the environment "voluntarily" in response to economic offers from presently regulated sources where it is cheaper and more efficient to control the unregulated sources. Such a regime gives flexibility for local adaptation, and it also creates incentives to pay more attention to environmental opportunities for those outside the present regulatory system. For example, service sector companies like delivery companies or Internet providers, who may not as yet have specific pollution control obligations, might be induced to come up with programs to reduce emissions elsewhere in the economy through a legal system that provides positive incentives for pollution reduction.

Similarly, it may not be possible to regulate consumers' life-styles through negative governmental orders that restrict choices without producing a political backlash, but when the same changes are led by positive incentives, they can prove acceptable. For example, experience shows that in many circumstances, paying commuters to ride the subway (through employer-provided subsidies) is more effective than trying to get them out of their cars through negative regulatory incentives, such as limiting parking spaces and raising tolls (as some states tried, unsuccessfully, under the Clean Air Act in the 1970s).

As we move toward an economy in which large industrial sources of pollution are less important, our best chance of effectively regulating agriculture, the service sector, and consumer products may lie in the "carrot" of voluntary tradeoffs between sectors rather than the "stick" of expanding command-and-control regulation. A refinery that has already controlled most of the sources of volatile organic compounds (vocs) within its boundaries that are easy and cheap to control may be able to achieve needed additional reductions more efficiently by paying a local dry cleaner to upgrade its machinery to reduce vocs, or by redesigning a consumer product to eliminate voc releases to the environment. The incentives to find innovative opportunities to reduce pollution—primarily from the multiplicity of pollution sources that are presently outside the existing command-and-control system—is one of the most attractive features of expanding the bubble concept.

The simple concept of allowing enforceable contracts to trade greater benefits for lesser ones is already included in a different but somewhat similar form in the "dual" system in Germany, or the "covenant" system in the Netherlands.[8] Similarly, the Clinton Administration's epa has experimented on a limited basis with Project XL, a program to permit polluters to comply with environmental requirements in ways that do not comply literally with statutory mandates where there is strong community support for the alternative approach. Like the bubble, these approaches leave the existing system of regulations in place but gradually transform it into a baseline against which more effective, custom-designed arrangements can be judged.

These two-tiered approaches can serve as the legal instruments that create incentives for companies to try new practices as the building blocks of transformation, as Charles Powers and Marian Chertow describe in chapter 1. As experience with the new approaches to assessing environmental performance grows, and confidence in them builds, further transformations of the legal system will become possible. This may frustrate some of the visionaries who believe that they see clearly the proper road ahead; but those who study the process of legal change know that a system which is a qualified success, entrenched, and politically controversial, as today's environmental law such, is unlikely to be overthrown by even the most promising of ideas. Abstract argument that there is a better way will not convince the skeptics; only verifiable experience can demonstrate that there really is great promise for further environmental improvements with legal technologies more advanced than bureaucratic command and control. We need to start small but think big, by beginning to implement changes that start out as incremental but can grow to transform radically the system itself.

From "Command and Control" to "Command and Covenant"

There is a strong, bipartisan consensus on the outlines of a set of next-generation environmental law and policy tools that meet the seemingly contradictory requirements of incremental changes with radical, transformative implications. The use of the existing system of regulations as a minimum benchmark with increased flexibility at all levels of implementation for those who can achieve equivalent or better environmental performance represents a point of departure on which many can agree.

This idea of flexible compliance with performance equivalent to or better than the existing system goes by a variety of names—alternative compliance, Project XL, the multimedia or risk bubble, "challenge regulation," or "hybrid regulation"—but the central concept can be thought of as a move from "command and control" to "command and covenant." The government still "commands," in the sense of identifying minimum acceptable levels of risk or environmental degradation. But rather than "controlling" how regulated parties achieve compliance with these environmental goals, implementing agencies (whether states, local districts, or individual factories) are empowered to design their own enforceable alternative compliance methods or covenants, provided that they demonstrate that the alternative achieves equivalent or better environmental performance. This approach essentially allows private parties to "contract around" inefficient government regulations by substituting a more efficient alternative for achieving an equivalent level of environmental performance.

The primary advantage of this system of flexible compliance is that it provides incentives for innovation and achieving environmental compliance at lesser cost. A portion of the cost savings should be shared with the public or the environment by requiring some increment of improved environmental performance rather than mere parity with the existing system. This "margin of advancement" would not only enhance the acceptability of flexible compliance but would also compensate for the risks of slippage and misstatement that inevitably accompany a new system. In addition to lower costs, however, experimentation with flexible compliance systems will provide valuable experience with more innovative techniques and build confidence that can be used to make the transition to the next stages of government policy.

A fundamental paradox affects alternative compliance schemes. On the one hand, the more broadly the measures of equivalent compliance are defined, the greater the opportunities for flexibility, improved performance, and cost savings. Thus, a comprehensive risk bubble that would permit one kind of environmental risk to be traded off against another would potentially provide the greatest benefit. On the other hand, the difficulties of measuring "equivalency" become greater the more broadly the concept is defined. Thus, it is inherently easier to equate one pound of sulfur dioxide released in one part of a factory with an equivalent pound of the identical chemical released somewhere else in the same facility. However, as common metrics are developed for comparison purposes, dissimilar environmental risks will be able to be traded off more broadly against one another.

In addition to these problems of comparing different environmental risks, alternative compliance systems inherently raise the problem of verifying that the substituted compliance is really equivalent. The solution might be "trust, but verify." Let those who benefit from more flexible compliance have the burden of clearly measuring and documenting that the alternative system delivers better results than the traditional mechanisms it has replaced, using government-approved validation mechanisms. Intentional disregard for the established standards should be punished harshly, and unintentional violations or failures to achieve compliance should result in environmental give-backs that more than make up for any increased pollution. Allowing compliance alternatives to the current system takes the present level of environmental protection as a given, and thus does little to correct for either over- or undercontrol reflected in the present level of environmental regulation. However, taking the present level of environmental protection as a starting point seems likely to be a crucial first step toward improved efficiency while maintaining credibility. Moreover, as our measures of environmental performance improve, broader definitions of risk trade-offs will be authorized, and more efficient reductions substituted for those that are unduly regulated by the present system. In this way, the excesses of the present system—in the direction of either overcontrol or undercontrol in a particular area—can be corrected by the incentives of market forces, while assuring the public that the environment will continue to be protected.

Planning Institutional Change

Today there is rapid innovation and experimentation in the environmental area, as many new approaches to implementing environmental programs are being tried and evaluated. In no small measure, this period of innovation and experimentation is being promoted by the movement toward devolution of power away from Washington and toward the states, but also to individual localities and facilities, which are given more flexibility to develop the means to achieve national environmental goals. In addition, increased globalization promotes borrowing of successful techniques across national borders. In a sense, the overarching philosophy of national goals and flexible local implementation is not new. In theory, most federal environmental statutes merely set minimum national standards. But in practice, in the past national regulations were usually drawn so tightly that relatively little flexibility for variation in implementation existed. As environmental law matures, those closer to problems—and closer to designing solutions—must be trusted with greater discretion to devise more innovative solutions. A command-and-covenant system creates increased flexibility and opportunities for innovation, while maintaining accountability.

Nature as Guide to Reforming Environmental Law

Environmental lawyers should be the first to hear the counsels of nature about how complex systems change over time. None of the many innovative ideas for reforming environmental law reviewed above is in and of itself the answer; there is no philosopher's stone or panacea that will solve all our problems with a single stroke. But even if there were, no complex system like the environmental law we have built over the last generation changes over night. Complex systems do not lurch forward; they adapt, blending present into future through subtle changes that transform what already exists by creating new relationships. Understanding that process is the key to shaping it.

Multimedia bubbles and command and covenant are the right strategy for the next generation of environmental law because they are the bridge between the present, with all its strengths and weaknesses, and the future, with all its promise and uncertainties.

Notes

1. EPA, *Environmental Investments: The Cost of a Clean Environment* (Washington, D.C.: Government Printing Office, 1990).

2. See Introduction, n. 6. See also E. Donald Elliott, "Foreword: A New Style of Ecological Thinking in Environmental Law," *Wake Forest Law Review* 1 (1991): 26.

3. Richard Stewart, "Economics, Environment, and the Limits of Legal Control," *Harvard Environmental Law Review* 1 (1985): 9.

4. See E. Donald Elliott, "Environmental Law at a Crossroad," *North Kentucky Law Review* 20 (1992): 1.

5. See *Chevron v. NRDC,* 467 U.S. 837 (1984).

6. The public should be better off in both an aggregate sense and distributionally. Specifically, no individual should be worse off.

7. Ronald Coase, "The Problem of Social Cost," *Journal of Law and Economics* 1 (1960): 3.

8. For more information on the "dual" system in Germany, see Bette K. Fishbein, *Germany, Garbage, and the Green Dot: Challenging the Throwaway Society* (New York: Inform, 1994). For more information on the Dutch "covenant" system, see Hans van Zijst, "A Change in the Culture," *Environmental Forum* 10, no. 3 (May–June 1993): 12–17.

three

Extending the Reach of Next-Generation Policy

Coexisting with the Car

Emil Frankel

Automobiles have shaped our values, our politics, and the patterns of our lives. They have provided us with highly prized mobility, greatly influenced where we live and work, and, by stimulating the construction of a vast system of roads and highways, encouraged a national pattern of low density land use and intensive suburban development. They have also had a profound impact on the air we breathe, the water we drink, and the land we use. Automobile engine exhaust pollutes the air of our cities and contributes "greenhouse gases" to the atmosphere. The construction and daily use of sixty thousand square miles of American roads have changed the drainage pattern of our land and damaged our wetlands. Yet the automobile remains our overwhelming choice for transportation. Ask for our help improving the environment, and we agree. Try to force us out of our car or change the way we use it and we shrug guiltily as we head for the driveway, car keys in hand.

First-generation environmental policies curbed some of the most severe impacts of a transportation system built around the automobile. Catalytic converters, vapor recovery systems on gasoline station pumps, new gasoline formulas—all helped to eliminate airborne lead and to cut in half the amount of carbon monoxide and volatile organic hydrocarbons released by motor vehicles during the twenty-five years after 1970. However, although we have been making progress, growth in automobile ownership and use has threatened these air quality gains. We cannot change this merely by passing new laws. Transportation policy is not determined by a few public officials. It reflects the sum of the decisions by businesses on where to locate and how to schedule work hours and deliveries and by all of us when we choose where to live or work, how to get from one place to the other, and how fast to drive while doing it.

While consumers were increasing their use of cars, manufacturers and retailers were adopting practices such as "just-in-time" inventory management that have increased the shipment of goods by trucks in ways not anticipated by first-generation policymakers. No method of transporting goods is as flexible or permits as much time precision as sending them by truck. Neither environmentalists nor planners of the Interstate Highway System foresaw the way truck traffic would come to dominate many highways. In Connecticut, for example, 95 percent of goods move by truck. Almost all freight moves by truck at some point. Many factories maintain such low inventory levels that trucks moving on highways have effectively become their warehouses. (See the section on logistics and distribution in chapter 5.)

First-generation policies also failed in part because they tried to change or dictate behavior. For example, considerable effort was placed on reducing the American preference for driving alone, particularly for the daily commute. One reason those policies failed is that homes and jobs continue to decentralize. The dispersion of people and work locales has accelerated over the past two or three decades to the point where more than 50 percent of the nations' commuters and more than 41 percent of the jobs are located in the suburbs. But this outward movement is just part of the problem. For most people and businesses, use of automobiles or trucks is a rational, commonsense decision. Other ways of traveling might get people or goods from point A to point B with far less pollution, but none affords the flexibility of the automobile.

Anthony Downs, a senior fellow at the Brookings Institution, observes, "Fundamentally, you can't get people out of their cars, because cars are a superior way to move."[1] We have complex travel patterns that become even more tangled as the two-earner family becomes the norm and as businesses decentralize. Driving her own car, the working mother leaves when she's ready, arrives when she wants (congestion permitting), and stops along the way to drop off the kids, retrieve the dry cleaning, or grab a meal—all in comfort, privacy, and, especially if parking is free, with the appearance of low cost. The attraction of your own car under these circumstances is hard to resist— and few do resist it. As Downs and others have pointed out, more Americans are driving and driving more. A Cato Institute study notes that from the mid-1970s to 1990, while U.S. population increased by 16 percent, the number of cars increased by 42 percent.

Next-generation policies need to take these realities of behavior into account, to "listen to the customers." Several first generation policies foundered because they did not. High-occupancy-vehicle (HOV), or diamond, lanes leading into or through city centers did not significantly reduce traffic congestion and miles traveled by motor vehicles partly because even with this inducement few people chose to car pool. Moreover, jobs were moving to the suburbs and more goods were moving within and between metropolitan areas by truck. Similarly, even in cities like San Francisco and Washington, D.C., where there have been major investments in new public transit systems, road traffic has continued to increase. New or improved transit systems can take pressure off highway systems in some densely populated regions and travel corridors. However, in most areas of the country downtowns are disappearing, while office parks and shopping districts thrive outside the urban area, where they are difficult to serve with conventional forms of public transit.

The speed limit and the gasoline tax provide other examples of how hard it is to change driver behavior. The 55 m.p.h. national speed limit reduced pollution and saved resources as well as lives. Moving the limit up to 65 dramatically increases the amount of volatile organic compounds huffed into the air by cars. Yet Americans literally drove over the 55 m.p.h. limit—all the while complaining about smog. Even before federal and state governments conceded defeat, raising speed limits in many areas, the lower limit was largely unenforceable and unenforced. Similarly the 4.3-cent increase in federal gasoline taxes in 1993 moved the United States in the direction of reducing some unnecessary driving, but when gasoline prices rose in early 1996, few politicians rose to defend the tax, even though there was little clamor among voters for its rescission.

Policymakers must recognize reality: Americans cannot be forced out of automobiles by regulation nor cajoled into using them differently by high-minded appeals to sacrifice personal convenience for the good of the whole. But at the same time, we cannot drive away and forget the problem. The key will be making car and truck travel pay for itself. When the full costs of pollution, congestion, and habitat destruction are factored into driving, incentives for change—in personal behavior, corporate transport decisions, and technologies—will be created.

It won't be easy. But it is not impossible. The following are some promising developments, techniques, and strategies:

Fuel technology and automobile design. Progress in these areas will come in two phases—an incremental improvement to reduce the current level of pollution and, later, more radical innovation to keep up with a growing population of people and vehicles. The first phase for fuel includes new formulas for gasoline and conversion to burning natural gas and propane. If, for example, all the taxis in New York City were converted to natural gas, that region's air emissions would be cut in half. The second phase will need greater technological sophistication and will most likely be a move from exploding natural fuel in a cylinder to electric power, possibly generated by hydrogen fuel cells, or hybrid engines that use different energy sources for different purposes.[2] In automobile design, initial development will be improvements in weight, streamlining, and tire resistance. General Motors' Ultralite prototype, for example, is about half the weight of a conventional family car, but it has zippy acceleration and plenty of interior space, is safe, and doesn't look revolutionary. In normal highway use, it travels fifty-six miles on a gallon of gasoline. To do better, we need a super-efficient car with a new shape, construction materials like carbon fiber, and some major change in the power system.

Many of the new automotive and fuel oil technologies now being investigated are expensive relative to the mass-produced, gas-powered automobile and the high-volume output of refineries. Incentives may be needed to facilitate the transition to new modes of transportation. Building cars with high-tech materials will mean developing new manufacturing techniques and purchasing new fabricating equipment. Producing cleaner fuels will mean, at the least, major alterations of refineries and, at the most advanced, the replacement of refineries with new production, distribution, and storage facilities. Ultimately, the cost will have to be passed to vehicle buyers and operators. But how much are we willing to pay for this? And will manufacturers be willing to take risks in a highly competitive market and make changes that raise prices?

Government can prod change. California and Massachusetts, for example, will require that 10 percent of all cars sold within their borders be "zero emission" vehicles by 2003. These requirements will likely induce the auto industry to refine their prototype electric cars and to invest in better batteries. But in the long term, technological innovation will result from a mix of regulation, government procurement and incentives, and the dynamics of the marketplace.

"Intelligent" transportation. Tollbooths that electronically deduct money from your account after you drive by are the visible edge of a still-infant "intelligent transportation system" (ITS). In the near future, sensors buried in the roads will relay various kinds of information to highway operations centers that can alert emergency vehicles, use electronic signs to advise drivers on road conditions or traffic delays, and continuously alter charges to reflect road congestion. Similar devices are now available to check the weight and registration of moving trucks, allowing most to bypass weigh stations.

Beyond smart roads and cars, these technologies can be used in transit to automate buses, making them respond to riders' demands and provide real-time schedule information to transit patrons. Ridesharing can also be automated by means of technologies that match drivers and passengers, conduct security checks to ensure safety, and electronically transfer money from a passenger's account to pay a driver for a ride. If done right, ITS increases information, assists in the implementation of pricing, and could revolutionize transportation in the next generation.

Mass transportation. The debate over mass transit has gone on for decades. Some assume that public transit systems are operated inefficiently and that privatization could solve the problems of increasing cost and declining ridership. Others argue that equity is the prime goal, that government-owned transit is a public service required to offer even highly unprofitable "social services" at "reasonable" fares. Nevertheless, as state and local transportation deficits grow, the quest to optimize transit services has resulted in various suggestions for their restructuring. One option for buses is to allow free competition among private jitneys who register schedules in advance and, at the same time, grant exclusive curb rights to publicly-operated transit. Another idea is to experiment selectively with restructuring services that could cover their costs, such as suburban commuter rails alongside heavily congested roads. This could allow strategic targeting of public resources to higher-cost transit operations serving low-income, elderly, and disabled individuals.

Massive shifts of individuals from automobiles to public transit are unlikely. Nor can we expect large growth in public expenditures on such systems. But selective public and private investments in both urban transit and high-speed intercity passenger rail systems can be justified in high-density transportation corridors where it might reduce some of the pressure to expand highway capacity. And more flexible transit alternatives can be used in less densely populated suburban areas to serve those who cannot, or choose not to, drive. In the next generation we should experiment with innovative options to provide mass transportation that maximizes efficiency and cost-effectiveness without sacrificing fairness.

Return of the railroad. In the past several years, public policies have removed many of the regulatory obstacles to the growth of means other than trucks to move freight. Trucks are likely to remain a critical way of assuring timely and flexible deliveries, particularly for shorter movements and for the shipment of high-value, low-volume goods. But thanks to deregulation, freight railroads are recovering and becoming more aggressive competitors. Some of the larger railroads are transforming into intermodal carriers, combining rail, truck, barge, and ship transportation. Although these modes themselves raise challenges to the environment, the intermodal approach holds the promise of more efficient utilization of resources in freight movements. This trend will become more important if some means is found to fully charge trucks for their impact on the highway system. In a genuinely nonsubsidized private marketplace in which companies can make decisions based on prices and quality of service, the more efficient—and less polluting—rail system will gain increased freight business.

Taxes and incentives. Taxing transportation-derived pollution along with the elimination of hidden subsidies for automobiles must be part of next-generation policy in the transportation sector. Charging emission taxes when a car is registered or inspected, which vary with pollution performance of each model of car, is the most direct way to get vehicles to pay for the harms they cause. New sensor technologies may even allow taxes to be imposed based on individual vehicle pollution. User charges that must be paid with every mile traveled also should be considered. In either case, appropriate emission and gasoline taxes will stimulate an effort to improve the fuels and design of cars to reduce emissions. The funds raised could be used to support the transition to cleaner and more efficient transportation systems and to mitigate equity concerns that might arise if poor people have to pay higher taxes.

"Cashing out" free employee parking would also remove a distortion that favors commuting by car. If companies paid employees a flat amount of cash and charged them for parking, some number of people would use public transit or car pooling. In any event, drivers would be put on equal footing with other commuters.

Putting a Price on It

As the *Economist* has put it, "Cars must be made expensive to use, rather than to own."[3] Charging for parking is just a start. If the price of driving accurately reflected its true cost, and if the price varied depending upon the time and nature of use, Americans would almost certainly use the highway system more efficiently, generating less pollution. For hundreds of millions of individual drivers, price signals, rather than exhortation or government mandates, are far more likely to succeed in changing behavior.

Privatizing the roads is another way to get there. Although highway-related taxes, imposed by governmental authority, will be a difficult political sell, private highway owners with franchises to manage and operate certain roads have no choice but to charge drivers fully to cover expenses and to earn a return on their investments. Governments will find it far easier to make these businesses pay fees for the environmental harms their systems cause. But for U.S. drivers who came of age with *free*ways, it seems heretical to contemplate turning over any part of the roads to entrepreneurs who will expect full payment in return for services rendered. After all, the 1956 Interstate Highway Act and subsequent bills under which the United States built some forty-one thousand miles of interstate highways specifically prohibited states from charging tolls on most roads built or improved with federal funds. However, while the demand for more roadway capacity keeps growing, the financial ability of governments to build and maintain them has declined. As drivers, we want better roads. As taxpayers, we have refused to pay more for them.

Toll roads—theoretically self-financing—are one solution if we accept that we cannot get people out of their cars or pay taxes for more roads. New private roads across the country from Colorado to Illinois, from Virginia to Florida are financed by user fees. The chance to alleviate congestion is diminished if toll rates are fixed rather than adjusted according to congestion and demand. In fact, most pricing schemes are upside down, with rush-hour commuters being rewarded with multiple-use discounts.

Turning over the job of building a toll road to a private company would pass the financing risk from taxpayers to investors who are more isolated from the political pressures that influence the decisions of state highway department personnel. Investors would fund a project only because they had carefully determined that it would attract paying customers. If it did not, they, not the taxpayers, would suffer the loss. Next, private owner/operators would have to incorporate in their prices all of their operating and capital costs plus a profit. There would be no free or subsidized ride for users, and thus no unfair competition for other means of transportation. If drivers had to pay the full cost of using highways, riding the train might become the cheaper, better way to get from city to city.

Charging for the use of the roads also opens the way to include in the fee the costs of mitigating the impact of motor vehicles on air quality, habitats, and other resources. Attractive as that sounds, it is admittedly far easier to contemplate than to execute. Although economists can roughly estimate the monetary cost of automobile-caused pollution, identifying the harms accurately and then allocating them to a mile driven on a specific road at a specific time is far more difficult. Among the problems are differences in perceptions of the value of clean air or the integrity of a wetland. Ultimately, however, setting and exacting a price is the only way to internalize the cost of pollution effectively. But fairness as well as good environmental policy dictate that we shift the cost of pollution harms from the general public to those individuals responsible for producing them. We expect citizens to take personal responsibility for so many other aspects of their lives, why not for their pollution?

If charges are varied by time of use and number of occupants per car, drivers will be rewarded or penalized for their choice to increase or reduce the impact of the highway system on the environment. If they have swallowed the bitter pill of paying for something that has been perceived as free, users ought to be able to digest a refinement: pricing that varies by time of day or by day of the week. Most of us understand why it's cheaper to call Mother in the evening or on the weekend. Moreover, variable pricing aligns the interest of the driving public and the private company operating a highway. We want less congestion. The entrepreneurial owners want to find ways to increase the capacity of a road or other facility at the least expense.

Will consumers (and voters) who now don't directly pay for using the highways be willing to do so if, in return, they can have a greater degree of certainty about how long the trip will take? Three states are trying to find out:

- Virginia, the first state to approve construction of a private toll road, did so to prevent a problem from developing. Running fourteen miles west from Washington's Dulles Airport and ending in bucolic countryside, the privately funded $325 million project is now open. Built with care for its impact on the land, the road has been supported by many environmentalists who look on it as a way to concentrate inevitable development along its path and thus take pressure off the surrounding rural, "hunt country" areas. Light use since it opened in 1995 has concerned some toll road proponents, but the new road is expected to prevent future congestion.[4]
- California is trying to solve a problem that already exists by approving four private toll roads, all aimed at relieving highway congestion. One probably will not be built. Just east of San Francisco-Oakland, it has encountered stiff opposition, particularly from supporters of BART, the area's mass transit system. The other three, all in southern California, have been generally welcomed and one project, the $125 million SR 91 Express Lanes, has been completed, to the great pleasure of its paying customers. The Express Lanes are ten miles of four-lane highway built in what would have been the median. Until the lanes opened, Californians from the inland bedroom communities crawled twice a day along this route on their way to and from the coastal freeway joining Los Angles and San Diego. Now a high-tech operations center continuously varies tolls from twenty-five cents to ten times that in order to keep traffic moving at a guaranteed fifty miles per hour.[5]
- Washington State officials have been trying to turn over existing highways, bridges, and related systems to private operators as part of a four-phase effort to modernize and privatize the roads and bridges in the Seattle-Tacoma area by 2010. In the proposed first phase, private investors will take over and upgrade the Tacoma Narrows Bridge as well as freeways in the Seattle area. Among other changes, single drivers will be able to buy their way onto lanes now reserved for buses and carpools. (The high-occupancy-vehicle designation, HOV, is being changed to HOT, high-occupancy or toll.) Later phases are supposed to include private funding and operation of 166 new miles of HOV lanes. However, after the first franchises were awarded, the Washington legislature gave way to public complaints and set new requirements that the

private operators might not be able (or wish) to meet. It is still not clear whether the Washington effort will get under way.

All of these projects are little more than experiments or efforts to get public approval to experiment. But at least the southern California and Virginia projects signal that some drivers can be convinced to pay tolls for better, higher-capacity roads that they would not approve as taxpayers. (It's not surprising that voters find it far more palatable to pay for a road after it is open than to approve taxes in return for politicians' promises to build it later.) The new projects also demonstrate that private capital can be accumulated for highways. Finally, they offer hope that drivers will adjust their behavior in response to price in much the same way as they choose to purchase or not purchase any other commodity. The dynamics of the private marketplace, the interplay of supply and demand, rather than prescriptive governmental regulation, will then influence when and how much we drive and, eventually, what cars we buy as well as where we live, work, and shop.

Privatization and pricing of the highways will be difficult. Besides such thorny problems as how to cost environmental damage, entrenched bureaucracies will resist change. Compromises have to be crafted between those whose immediate, uncritical reaction is to laud the efficiencies and innovations of private enterprise and those who are just as quick and reactive in their condemnation of what they see as entrepreneurial greed and lack of social concern. Public discussion has to be led through debates about turning over existing highways that are seen by drivers as "free" or about the social equity of "regressive regulation" that appears to bar the poor from the highways. But without radical changes, within a few years many of our highways and roads will fail as dependable and predictable means of moving people or goods.

Where Do We Go from Here?

The dominance of the automobile in post–World War II America has been a critical factor in shaping a society of mobility and independence and in making a livable environment accessible to many people. We can't "regulate away" the automobile age, nor should we. But we can influence personal behavior and the use of the transportation system in ways that mitigate its most serious environmental consequences. We can introduce the dynamics and discipline of the private marketplace into what has been a sector almost totally dominated by public and quasipublic agencies. We can find ways to pass the cost of environmental damage to the drivers who create it. And we can support and try to stimulate innovations in the management of transportation systems and in fuel and vehicle technologies.

Most of us agree that we need a more efficient and less wasteful transportation system that balances environmental impact and personal mobility. We all want to reduce pollution, eliminate congestion on our highways and streets, increase intermodal freight activities, and, where possible, find practical and efficient mass transit alternatives to the automobile. But agreeing on how to accomplish all this will require leadership and compromise. There are no grand solutions, no painless rescues that will overcome persistent and strong public opposition to increasing the cost of automobile use. But an era of limited public financial capacity provides an opening for full-cost pricing through a mixture of means. Technology innovation can also help. In the end, educating the public that driving is not "free" is the key. People should pay for the pollution or congestion they create. When this principle is established, we will have the foundation for an environmentally sound transportation system.

Notes

1. Anthony Downs, quoted in Ben Wildavsky, "King of the Commute," *National Journal*, 20 Jan. 1996: 115.

2. The "tailpipe" technologies, such as catalytic converters and new fuel formulas, have done a lot to clean up the air at street level. But they haven't reduced carbon dioxide generated by the motor vehicle engines that contribute to global climate change. Either a new energy source that does not use hydrocarbon combustion or enhanced vehicle fuel efficiencies will eliminate the "greenhouse gases."

3. "Living with the Car," *Economist*, 22 June 1996: special section, 17.

4. The operators of the Dulles Greenway are currently losing money but are looking at differential pricing as a way to increase revenue. See *Innovation Briefs* 7, no. 6 (December 1996), published by Urban Mobility Corp., New York, edited by C. Kenneth Orski.

5. The success of SR 91 is reported on in *Innovation Briefs* 7, no. 1 (February 1996): 1.

thirteen

Environmental Protection from Farm to Market

C. Ford Runge

Elsewhere in this book, authors advance next-generation environmental policies for many sectors of the economy. But agriculture is different. It never had coherent first-generation environmental protection programs. Heavy federal intervention to control crop production and subsidize prices makes agriculture one of the most regulated sectors in the U.S. economy. At the same time, command-and-control environmental policies have never been applied to crops and livestock. Not one of the expensive schemes that have characterized agricultural policy since the 1930s has adequately controlled water pollution, pesticide overuse, or species losses.

It is important to examine the history of farm policy to see how linkages can be established to benefit both agriculture and the environment. The pristine wilderness and unbroken prairies that confronted early settlers seemed in need not of protection but of taming and exploitation. Through two centuries, America's land and water were divided, redirected, and aggressively used. Little environmental protection was asked of America's farmers in return for their bounty until the 1930s and the days of the Dust Bowl. At that point, the federal government intervened. Programs were introduced to prevent oversupply of farm products by forcing farmers to stop production on acreage that was then placed in "conservation reserves." Thus began a sixty-year reign of government intervention that retired land primarily to prevent the accumulation of surpluses and only secondarily to conserve soil resources.[1] When surpluses declined and agricultural prices strengthened, the soil conservation programs were usually abandoned.

In the 1950s, for example, a federal "Soil Bank" paid volunteering farmers to take land out of active cropping and plant grasses and trees along hillsides, watercourses, and field edges. By 1972, however, the Soil Bank, which had held tens of millions of acres at its peak, was almost empty. A major export boom and fencerow-to-fencerow cropping had driven land values steadily higher, encouraging farmers to put all available land back into production. Then, in the mid-1980s, the boom turned to bust. Commodity prices fell. Surpluses returned. With the added stimulus of a surge of environmental consciousness, a new land conservation initiative—the Conservation Reserve Program (CRP)—got under way in 1985. Like the Soil Bank, the CRP paid farmers to take land out of active cropping, this time over a contract period of ten years. But by the mid-1990s, supply shortages and high grain prices had once more encouraged many farmers to abandon conservation and resume production.

In 1996, a new farm bill, the Federal Agricultural Improvement Act (FAIR) of 1996, yielded some important breaks with the past. But other than voluntary conservation payment schemes including a reauthorized CRP, no significant environmental controls have been placed on farm practices even where agricultural activities are a primary cause of pollution problems. Those restrictions that do apply (for example, in the case of wetlands protection) remain hitched as in the past to withholding subsidies rather than assessing penalties. As a result, they are least effective when prices are strongest and subsidies lowest—precisely when farmers produce most intensively.

Other policy responses to environmental pollution in agriculture have also generally been "carrots" rather than "sticks." These have included a variety of cost-sharing schemes administered through federal and state agencies under which farmers are paid for constructing ponds, drainage areas, and similar projects. Instead of a "polluter pays" principle, under our agro-environmental policy the farmer must be bribed not to pollute. In some cases, agricultural pollution is simply ignored. The Clean Water Act, for example, largely exempts agricultural pollution from regulation as "nonpoint" sources—in particular, runoff from most fields—even though such contamination accounts for enormous damage to the environment.[2] As a result, fewer than 10,000 out of 1.1 million American farms are subject to the Clean Water Act.

The absence of longstanding environmental policies in agriculture nevertheless creates opportunities for:

- Selective deregulation of farm policies to harness market forces that will reduce farm-level incentives for environmentally harmful activities while tightening controls on the most environmentally sensitive agricultural practices, such as excessive pesticide use and cultivation of riparian areas and other vulnerable lands;
- new technologies that economize on the use of polluting fertilizers and pesticides while reducing farm costs;
- transferring substantial authority for implementation of environmental conservation to states and localities, leaving federal policy to set broad standards and provide needed investment funding;
- better information and data collection on landscape, soil, and watershed characteristics which will allow policymakers to target lands where intervention will be most cost-effective.

Seizing these opportunities for change could increase farm profits while reducing the environmental impact of agriculture. Agricultural production in the United States is a vast enterprise, from field crops such as wheat, corn, and soybeans to fruit and vegetable production, from cattle rearing on open range to intensive poultry and hog production in large confinement facilities. Agricultural production accounts for over half the nation's 2.3 billion acres of land in use, including 460 million acres of cropland and 591 million acres of grassland, pasture, and range. With cropland concentrated in the central United States and grassland, pasture, and range concentrated in the arid West, agricultural activity is highly differentiated by region, with substantial variations in moisture, soil fertility, and topography. Different areas respond quite differently to the possible stresses of agricultural production. Therefore, one-size-fits-all policies simply are not practical.[3]

Although the total cropland acreage in the United States has remained roughly stable since World War II, agricultural output has increased dramatically as a result of the increasing intensity with which land is used. Methods that farmers have used to boost production include hybrid seed varieties, improved mechanical harvesting and cultivation, supplementation of soil nutrients with fertilizers and chemicals, the application of pesticides, and the expansion of irrigation. From 1949 to 1994, corn yields per harvested acre nearly quadrupled, while the output of wheat rose more than 150 percent. This effort to coax more and more product from the soil has succeeded, but it has had a damaging impact on water quality and quantity, raised questions of food quality and safety, and reduced the diversity of plant and animal species. At the same time, an explosive expansion of intensive livestock-rearing and animal-feeding facilities has created waste management challenges that equal and often far exceed those facing municipalities and other industries.

Throughout the postwar period, federal government efforts aimed at managing supply and demand have almost incidentally allowed some land to lie fallow for varying periods of time. With the changes resulting from the 1996 Farm Bill, however, such federal mandates will cease and land use decisions will be almost solely driven by market conditions. The change could not have come at a more critical time. As the Northern Hemisphere entered the 1996 growing season, supplies of grains and oilseeds were at their lowest levels in a half-century and rising grain prices were setting off a slaughter of livestock herds that will eventually drive up meat prices. As a result, market forces are now creating powerful pressures to raise output.

In the face of these pressures, and in the absence of significant constraints, mitigating the adverse environmental effects of agriculture depends significantly on the willingness and desire of individual farmers to conserve and protect vulnerable croplands, watersheds, and threatened species and habitats. Most farmers and others in agriculture endorse good stewardship and a clean environment achieved through vaguely defined "best management practices." But they are likely to do what is environmentally right only when it coincides with their economic interests. When farmers are asked to take land out of production or undertake other environmental investments that do not contribute to their short- or long-term profits, they often balk. It is on these points of divergence between individual and societal interest that next-generation environmental policy must focus.

Agricultural Environmental Impacts

The Office of Technology Assessment (OTA) identified three factors that determine the scope and severity of agricultural impacts on environmental quality. First, since agricultural production covers over half of the nation's lands, its impacts are widespread but concentrated in certain vulnerable areas. Second, some areas can adapt to and absorb the impacts of agricultural production better than others, making them less vulnerable. Third, agriculture technologies and agricultural policies have not, until recently, been adapted to these differences in vulnerability. In assessing the agro-environmental problems about which the most is known, three vulnerabilities stand out: water quality and quantity, wildlife habitat, and soil quality. These vulnerabilities pose challenges for water use and management, raise questions of food safety, pose serious threats to biodiversity, and have resulted in dramatic changes in land use.

Water. Problems with water are the most pervasive agriculturally induced environmental harm. The sheer quantity of water used by agriculture is, by itself, a significant issue. Intensive irrigation has drawn down the huge Ogallala aquifer stretching across Kansas, Nebraska, and Colorado and posed the possibility of significant future shortages and reduced productivity. Further west, U.S. Bureau of Reclamation projects have dammed river systems such as the Colorado and Columbia, and diverted billions of acre feet of water to agricultural irrigation at prices that average one-tenth those charged to nonagricultural uses. These water subsidies distort farm economics, undermine price signals reflecting the true scarcity value of water, aggravate drought conditions in urban areas, and deprive wildlife of vital water supplies.

Beyond the issue of excessive water use, farming activities are responsible for serious water quality problems across the country. Poorly managed agriculture is the primary cause of this impairment. Soil erosion, land conversion, pesticides, and animal wastes all induce changes in water, sometimes with dramatic effects on plants, fish, and drinking water quality. Groundwater, on which half of the U.S. population and most of U.S. rural communities depend, is especially susceptible to nitrate contamination from inorganic fertilizer and manure. Surface water is vulnerable to easily dissolved fertilizer and pesticide residues—nitrates, phosphorus, or herbicides such as atrazine—that then concentrate in streams, rivers, and lakes.

The surface water problem is most dramatic in the Corn Belt, where pollutants collect in streams and rivers that flow southward, contaminating the waters of neighboring states. Most of the phosphorus in rivers in eight midwestern states comes from other states. Sixteen states receive more than half of their concentrations of atrazine via upstream watershed flows.[4] Hundreds of thousands of tons of agricultural contaminants finally reach Louisiana's Gulf Coast estuaries, contributing to an offshore "dead" zone by stimulating algae growth that draws oxygen below the survival level for shellfish and other organisms. Some 80 percent of the nitrogen delivered to the Gulf originates more than a thousand miles upstream above the confluence of the Ohio and Mississippi Rivers—almost all of it dissolved in runoff from cropland.[5]

Given this flow of contamination, targeting water quality interventions upstream can prevent pollution downstream, sometimes much more cost-effectively than paying for pollution reductions downstream (at the "end of the pipe," in industrial terminology). However, although federal, state, and local governments have invested billions of dollars in municipal water treatment, the quality of water in many lakes and rivers continues to decline.

Food safety. Water problems may be widely recognized by experts, but food safety issues are often closer to the concerns of average consumers. Although an extraordinary food network makes the variety and prices of American produce the envy of the rest of the world, long marketing channels increase the need for preservatives and other interventions designed to reduce the chances for spoilage or contamination. Occasionally, outbreaks of food-borne illness or disease alarm the public, leading to calls for better oversight of food safety in general. In recent years consumers have focused primarily on pesticides—insecticides, herbicides, and fungicides—applied to food at the farm and potentially remaining as traces or residues in or on food when it arrives in the kitchen.

The Environmental Protection Agency "registers" pesticides and sets "tolerances" or limits for the amount of pesticide residue "allowed in or on a raw agricultural commodity and, in appropriate cases, on processed foods." But the agency's tolerance-setting and pesticide-registration processes have been widely criticized. The review process has been slow and considers pesticides one by one, with insufficient attention to the cumulative health impact of pesticides across the full dietary spectrum and little regard to what substitutes might come into the market if a product were banned or restricted. By focusing on previously unregistered products, the system has retarded the development of newer, more benign chemicals but allowed continued use of older chemicals. Another complaint reflects the lack of a systematic process for selecting which chemicals to review. Critics have charged that the EPA should target its energies on the ten compounds in fifteen foods with which about 80 percent of cancer risk is associated. Most recently, criticism has focused on a lack of attention to the interaction of various chemical risks such as the impacts of pesticides on infants and children, and on risks to the human reproduction and immune systems.[6] A 1993 National Research Council report, for example, found that infants and children, because of their developmental stages and their diets, differ "both qualitatively and quantitatively from adults in their exposure to pesticide residues in food."[7]

The EPA has recently started to require pesticide manufacturers to screen their products more carefully to determine their effect on both the human immune and endocrine systems. In July 1996, various factions in Congress agreed, after a decade of conflict, to overhaul the standard-setting process for pesticides. The resulting legislation streamlined the approval process for newer, safer pesticides, targeted children for protection, and required more pesticide information on food products.

The result may be less chemical-intensive food production. But American agriculture remains heavily dependent on these products. In the early 1990s, 368 million pounds of active herbicide ingredients, 51 million pounds of insecticides, and 33 million pounds of fungicides were used on major U.S. crops. Between 1964 and 1991, U.S. farmers' use of herbicides on corn grew from 26 million pounds to 210 million pounds of active ingredients. On soybeans, the amount used rose from 4 to 70 million pounds.[8] Part of the increase reflects a rise in planted acres, but most of the increase resulted from treating more fields with more chemicals. Still, there is opportunity at the farm level for the substitution of other production methods for pesticides. These include both crop rotations and integrated pest and crop management approaches.[9] Although these methods are still under development, they can be augmented and supported through a variety of changes in farm incentives. More use of organic product labels, for instance, will increase consumer choice and provide rewards for those who reduce their dependence on chemicals.

Biodiversity. Pressures on the agricultural landscape from increasing export and domestic market demands threaten both animal and plant species biodiversity. Conversion of many grasslands and wetlands to croplands, increased field sizes, reduced crop diversity, elimination of woodlands and field edges, reduced crop rotations, and increased fertilizer and pesticide use have had dramatic impacts on animal and plant populations, even among species well adapted to agricultural land uses, such as cottontail rabbits, quail, and pheasants. As all of the native tall-grass and much of the short-grass prairie was plowed under, the impact on species that were dependent on grasslands was dramatic. At least 55 grassland birds and other animals are listed as threatened or endangered, and 728 more are candidates for listing. Even the adaptable and prolific cottontail is under pressure. The number of eastern cottontail rabbits in eastern Illinois, for example, declined by 40 percent between 1956 and 1989.[10]

The Conservation Reserve Program, despite its weaknesses, demonstrates the benefits to wildlife when fields are converted from row crops to grass and woodland cover. The regions with the highest levels of CRP acreage, the Midwest and Great Plains, showed the most dramatic improvements in a variety of nesting pheasants and other animals. One CRP monitoring study extending over thirty states in 1988 found significant improvements in the level and suitability of habitat for a variety of game and nongame species. Another study found that several species of grassland birds that had declined dramatically in number from 1966 to 1990 were once again common in fields lying fallow in the CRP program.[11]

A nonadmirer of birds or cottontails might ask whether their survival is necessary or even important. The answer is that we do not know, although their consumption of a wide range of insects and plants harmful to commercial crops is an important contribution to productivity. In the case of plant diversity, we have more compelling experience. Researchers have determined that narrowing the genetic base through systematic elimination of plant varieties can have catastrophic consequences by creating susceptibility to a variety of plant diseases. In 1970, for example, southern corn leaf blight was spreading rapidly through much of the U.S. corn crop and was turned back only through the use of new corn varieties that had been held in storage by seed companies.

Agriculturalists have become increasingly aware of the need to conserve a diverse store of germ plasm by preserving stretches of native prairie as well as maintaining seed industry gene banks. The capacity of market forces alone to preserve this diversity is questionable. As Donald Duvick, formerly chief plant breeder for Pioneer Hybrid, International, the dominant seed corn company, noted, "The seed industry cannot finance such conservation nor should it, for germ plasm conservation is a public good and should be supported by the public. But seed companies can and should work together to provide support for the cause through their contacts with governments and policymakers, as well as through public pronouncements and other efforts to inform the public."[12]

Changing land use patterns. A final issue involves urban populations at the agricultural/urban fringe. As urban areas expand into productive farmlands, one set of pressures on habitats and biodiversity gives way to another—wholesale conversion of productive farmland and potentially valuable wetlands to urban uses, reducing the land available for crops and livestock and destroying plant and animal habitat. For example, in the Central Valley of California, the leading agricultural producing area by value in the nation, low-density urban sprawl could consume more than one million acres of farmland by 2040.

Throughout the debate leading up to the 1996 Farm Bill, policy analysts and most farmers generally agreed that federal agricultural policies should be changed. The issue was how and when. The eventual compromise reflects the analysts' desire for a more "market-oriented" policy, but still retains generous income transfers to farmers, although these are, in theory, to be phased out over seven years. On the environmental front, in addition to reauthorizing the CRP, the legislation provides new programs of cost sharing and direct grants to farmers for environmental improvements, especially in the area of livestock and animal waste management. Although it retains conservation compliance requirements affecting wetlands and highly erodible land, the new bill continues the tradition of carrots rather than sticks. But as the seven years pass, questions will arise concerning the efficacy of provisions that condition eligibility for subsidies on meeting conservation requirements. If the subsidy is gone, what incentives remain? Market mechanisms? If so, what is the potential for more market orientation in agro-environmental policies?

Experiments with market and quasimarket incentive mechanisms such as taxes on the nitrogen content of fertilizer have been undertaken in agriculture, but only to a very limited extent. There is no reason, in principle, why taxes cannot be used to encourage or discourage rates of use and alterations of the particular environmental characteristics of agricultural chemicals including water-solubility and toxicity. Iowa and a few other states have tried applying taxes to the use of fertilizers to encourage more careful use. So far, however, the taxes have not been large enough to offset the economic benefits to farmers of heavy fertilizer use. Most research suggests that changing the amount of fertilizer used on crops such as corn will require tax levels several orders of magnitude larger than those that have been tried.

The EPA has tried schemes to limit airborne pollution through programs in which participants trade "rights to pollute." Companies that emit less than a given target of pollutants can sell their unused rights to others (see chapter 7). In the Fox River area of Wisconsin and the Tar and Pamlico River area of North Carolina, agriculturalists are trying to adopt this approach through nutrient trading. In these programs, the objective is to establish and maintain an overall level of fertilizer use in a geographic area by allowing those who reduce their use below a target level to sell their rights to those who would otherwise exceed the target. Sellers are compensated for producing less; buyers are penalized and lose some, if not all, of the economic benefits of excess use of supplements.[13]

Given agriculture's huge contribution to nitrogen and other nutrient pollution and the sector's current rights to pollute, there is a large potential to lower land and water contamination through trading of pollution rights. Although this approach requires careful implementation to avoid simply "paying the polluter," it responds to the problem at the source, rather than at the end of the pipe. Although tertiary treatment of nutrients costs billions of dollars, adopting conservation tillage techniques on farms to reduce nutrient use and runoff often actually saves money for the farmer as well as the downstream public.[14]

Tax incentives and other market-oriented measures can also be used to stimulate and further expand a growing body of environmental technologies in agriculture. A promising set of these new approaches, known popularly as "precision agriculture," can be thought of as applying seeds, fertilizers, and farm chemicals at variable rates across different fields depending on careful evaluation and mapping techniques. These technologies are based on yield monitors, computer software, and variable-rate application and planting equipment, all of which represent large initial investments. As one observer notes, "Tax incentives or other cost-share programs could help tremendously to help offset costs and promote environmental advantage, particularly where water quality is at high risk and the local cropping and economic conditions are preventing initial investments from occurring, and the technology from being adopted."[15]

Despite these possibilities, market-based approaches will require clearer definitions of property rights. If it is clear that farmers have rights to nutrient use and loadings only up to certain thresholds, then market mechanisms will only encourage reductions below the point at which taxes or penalties are applied. All markets are imbedded in a structure of rules or performance standards that define trading limits and possibilities and are part of the successful market mechanism itself. In many cases in agriculture and elsewhere, property rights cannot be adequately defined, or the costs of doing so are prohibitive. For example, what rights do farmers have to reduce the populations of native wild grasses? And who should decide this?

As noted earlier, federal policies intended to impact agro-environmental issues have nearly always missed their targets because they have been hitched to other policies or objectives. The General Accounting Office, for example, examined what would have been the effect if the CRP had been carefully targeted on highly erodible lands rather than on trying to limit crop production by retiring higher-cost lands. The office concluded that all of the water quality benefits obtained over ten years could have been achieved by enrolling and paying for 6 million acres of buffer zones or "filter-strips" along streams and watersheds instead of the 36.4 million acres actually retired.[16] (It is still unclear whether the reauthorized, scaled-down 1996 version of the CRP will adopt such targeting mechanisms.) A second example is provided by efforts to deny federal subsidies to farmers who violate very modest restrictions on land use in environmentally vulnerable areas. Although these programs were rare examples of penalties in agro-environmental policy, they were tied to commodity subsidies that fell when market prices rose. Hence, in periods of strong market prices, additional profits dwarf the penalties for noncompliance.

An alternative to these self-contradictory policies is to set measurable and enforceable goals for agro-environmental improvements that are not tied to farm subsidies or surpluses. These programs should not be one-size-fits-all edicts, but should be adjusted as needed at the farm, county, state, regional, and national levels. They should be scientific in nature, focusing, for example, on measurable nutrient balances rather than qualitative and unmeasurable "best management practices." Farm conservation plans could be expanded to include not only target nutrient balances but also acceptable levels of pesticide use, and "credits" for nitrogen based on the farm's specific characteristics, such as soil conditions and erosion potential. (Improved crediting of nitrogen alone could reduce fertilizer use by as much as 40 percent, saving farmers several hundred million dollars per year.) At the county level, local officials could monitor overall nutrient balances and pesticide use levels and set goals for reductions in the deposit of waste materials into critical watersheds. These goals could then be aggregated once more at the state level, with the counties rewarded by the state (and the state by the federal government) for overachievement and penalized through fees or other assessments for underachievement.

Wisconsin provides some insight into the process of such state-level approaches and standard setting. As part of its groundwater protection strategy, it established rules to restrict the application rate of atrazine, a corn herbicide, focusing on areas where monitoring and assessment indicated that risks to water and the local populations were highest. The restrictions, which considerably exceeded federal standards, produced significant reductions in both the extent and intensity of atrazine use. As applications of atrazine fell, use of other herbicides such as cyanazine increased, a substitution anticipated and approved due to the lower mobility and persistence of these alternative chemicals. Many farmers, however, did not know the rules, limiting their desire to lower pesticide use or to adopt less troublesome substitutes. Officials would have improved results with better communications and more training for farmers about the alternatives available. Moreover, the restrictions were reactive rather than preventative. They were implemented only after serious groundwater problems had emerged.[17]

Policymakers also need to focus on ways to harness and stimulate environmental technology, already one of the most dynamic areas of innovation in the U.S. private sector. Three areas of opportunity in agricultural management technology stand out. Computerized land use records and better soil-mapping technologies would make managing public resources much easier by allowing public agencies to target their attention and resources where they are needed most or could make the biggest environmental quality difference. In Minnesota, for example, officials are using computer-generated soil maps as part of a program in which farmers are paid to change the way they use their land or, in some cases, not use it.[18]

A second new technology area is precision agriculture, or site-specific farm management. Among approaches that are both profitable and environmentally beneficial are soil nutrient testing and conservation tillage. In a case study of Pennsylvania farmers, nutrient testing reduced fertilizer applications by about one-third, a cost savings of $3.70 to $13.50 per acre. Conservation tillage systems, in which plant residues are left on the surface of fields rather than cultivated back into the ground, reduce erosion and conserve fuel, labor, and machinery. Such systems now account for nearly 40 percent of planted acres. The use of these techniques ought to be broadened, perhaps through integration into the farm-level conservation plans.

Integrated pest management, using beneficial insects and pest-resistant plant varieties to augment pesticides, represents another area of recent technological progress. Researchers have determined that the new approaches to pest management generally improve profits while lowering the use of pesticides—an important contribution to food quality and safety. These techniques might be especially important from the point of view of public health in urban/agricultural fringe areas such as California's Central Valley.

All of these technological innovations are already being adopted and promoted largely because they lower farm-level costs and only secondarily because of the attendant environmental benefits. Again, the central challenge for next-generation agro-environmental policymakers is to find ways to promote these new approaches to farming whenever the sum of private and social benefits exceeds the costs. A variety of additional incentives might be developed. It is, however, virtually impossible to devise optimal on-the-ground policies in Washington, or even in state capitals. Micromanaging farm-level behavior is almost certain to be counterproductive. A decentralized approach, relying on economic incentives, is necessary.

Yet, given the magnitude of the challenge, a mix of rewards and penalties will be required. Decentralization should not, moreover, be an excuse for deregulation that allows agricultural pollution spillovers or below-market-price consumption of public resources (like water) by farmers or ranchers to continue or even expand. A mix of federal, state, and local action and funds will have to be used. Responsibility must be shared across all levels of government and the private sector for developing new agro-environmental protection tools, ensuring that policies target the most critical public health and ecological harms, and that the resources committed to environmental protection are used effectively.

Next-generation agro-environmental policies should turn on five key ideas:

- clear, accountable standard setting at each level of administrative enforcement—from the farm to county, state, and federal governments;
- maximum use of market-based incentives to encourage the adoption of efficient farm practices and environment-improving technologies;
- an appropriate structure of financial rewards and administrative and financial penalties designed to close the gap between individual farmer interests and those of the broader community, encourage farm-level environmental progress, and discourage "backsliding";
- flexibility in the adoption of farm-level conservation and improvement plans to reflect local and regional variations in agro-environmental problems;
- conversion of federal funding for agriculture from price supports to environmental support, with responsibility for implementation passed down to the states and localities most directly affected.

These elements could all be captured with a "negative pollution tax" implemented at the local, state, or federal level.[19] For many years, some economists have promoted a negative income tax to replace current welfare programs: if family income is less than a given amount, there is no tax on its income. Instead, the family becomes eligible for payments intended to raise it up to a given level of annual income. The further below the threshold, the greater the payments. In the case of agricultural pollution, we propose a similar approach defined in terms not of income but of threshold levels of pollution from nutrients or herbicides, for instance, as determined by monitoring and evaluation (see fig. 13.1). Administrators would set a two-level threshold for individual farms. One (T-max) would mark the maximum acceptable usage limit of, say, a nutrient or pesticide. The other (T-min) would mark the desirable or target level. Each level would depend on the vulnerability of the particular area to the pollutant. A farm with accumulated usage above T-max would be fined or penalized. Below that, all farms would be taxed a decreasing amount as accumulations declined to the T-min point. Below T-min, farmers would be rewarded for environmental "affirmative action," by reduced taxes or even subsidies.

In addition to requiring clear and accountable standard setting, the refunds available for reducing pollution under the negative pollution tax program could be important incentives directed to new technologies such as precision farming

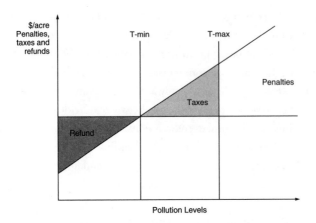

13.1 Diagramatic Representation of a Negative Pollution Tax

or integrated pest management. For example, state or federal agencies could agree to match a refund of, say, ten dollars per acre with an additional ten dollars if all of the money is used to convert cropland to a precision management program or to soil nutrient testing. Alternatively, if a particularly vulnerable land area at the urban fringe ought to be converted from cropland to rotational grazing, a refund could be supplemented with an additional payment to purchase a permanent easement for cropping rights while leaving the acreage for other agricultural uses. Finally, once the target range is established for particular land parcels, trading in pollution allowances would be permitted. A farm with a low level of accumulated nutrients might sell or trade its rights to refunds to another farm between T-min and T-max. The second farm could then reduce its tax burden and more easily implement conservation technologies. This added flexibility would enhance the capacity to target interventions while responding to different conditions on different lands.

Such a scheme maximizes the use of market incentives while retaining a set of agro-environmental targets. It could also make maximum use of current information technology to target land uses. And it would be at least partially self-financing, since revenues derived from the taxes paid could be used for refunds and subsidies. Clear priorities would be required at each level of administration, together with a willingness to acknowledge the need for carrots and sticks and government enforcement of the market-driven environmental protection regime.

In the final analysis, agriculture and the environment are linked by the choices, actions, and judgments we make as we take from nature to produce food and fiber. The history of civilization is replete with examples of fertile soils becoming unproductive wastelands through inattention or misconceived policies. We now have the opportunity to use science, technology, markets, and information to manage the agriculture-environment linkage, sustain agricultural productivity, and promote environmental quality for generations to come. But doing so will depend on changing our current choices, actions, and judgments.

Notes

1. Willard W. Cochrane and C. Ford Runge, *Reforming Farm Policy: Toward a National Agenda* (Ames: Iowa State University Press, 1992).

2. U.S. Department of Agriculture, Economic Research Service, *Agricultural Resources and Environmental Indicators: Agricultural Handbook 705* (Washington, D.C.: Government Printing Office, December 1994).

3. U.S. Congress, Office of Technology Assessment, *Targeting Environmental Priorities in Agriculture: Reforming Program Strategies* (Washington, D.C.: Government Printing Office, October 1995).

4. Richard A. Smith, Richard B. Alexander, and Gregory E. Schwarz, *Quantifying Fluvial Interstate Pollution Transfers* (Reston, Va.: U.S. Geological Survey, 1996).

5. Richard B. Alexander, Richard A. Smith, and Gregory E. Schwarz, "The Regional Transport of Point and Nonpoint-Source Nitrogen to the Gulf of Mexico," in *Proceedings of the Gulf of Mexico Hypoxia Management Conference, December 5-6, 1995, Kenner, Louisiana* (Reston, Va.: U.S. Geological Survey, 5 March 1996).

6. John Wargo, *Our Children's Toxic Legacy* (New Haven: Yale University Press, 1996).

7. National Research Council (NRC), Committee on Pesticides in the Diets of Infants and Children, Board on Agriculture and Board on Environmental Studies and Toxicology, and Commission on Life Sciences, *Pesticides in the Diets of Infants and Children* (Washington, D.C.: National Academy Press, 1993).

8. Gerald Wittaker, Biing-Hwan Lin, and Utpal Vasavada, "Restricting Pesticide Use: The Impact of Profitability by Farm Size," *Journal of Agricultural and Applied Economics* 27, no. 2 (December 1995): 352–62.

9. M. A. Altieri et al., *Agroecology: The Science of Sustainable Agriculture* (Boulder, Colo.: Westview Press, 1995).

10. U.S. Department of the Interior, National Biological Service, *Agricultural Practices, Farm Policy, and the Conservation of Biological Diversity,* by Philip W. Gerard, Bio-

logical Science Report 4 (Washington, D.C.: Government Printing Office, June 1995).

11. D. H. Johnson and M. D. Schwartz, "The Conservation Reserve Program and Grassland Birds," *Conservation Biology* 7, no. 4 (1993): 934–37.

12. D. N. Duvick, "Biology, Society, and Food Production: New Concepts, Old Realities," unpublished manuscript, March 5, 1996.

13. Note that the Tar-Pamlico trading scheme involves not discharges among farmers but indirect point-nonpoint trading of discharges. It allows municipalities to contribute to an existing agricultural best-management-practices cost-sharing fund to achieve equivalent or greater discharge reductions by farmers.

14. Paul Faeth, *Make It or Break It: Sustainability and the U.S. Agricultural Sector* (Washington, D.C.: World Resources Institute, 1996).

15. Richard M. Vanden Heuvel, "The Promise of Precision Agriculture," *Journal of Soil and Water Conservation* 51, no. 1 (January–February 1996): 38–40.

16. U.S. General Accounting Office, "Conservation Reserve Program: Alternatives Are Available for Managing Environmentally Sensitive Cropland," report to the Committee on Agriculture, Nutrition and Forestry, GAO/RCED-95-42, February 1995.

17. Steven A. Wolf and Peter J. Nowak, "A Regulatory Approach to Atazine Management: Evalutation of Wisconsin's Groundwater Protection Strategy," *Journal of Soil and Water Conservation* 51, no. 1 (January–February 1996): 94–100.

18. G. A. Larson, G. Roloff, and W. E. Larson, "A New Approach to Marginal Agricultural Land Classification," *Journal of Soil and Water Conservation* 43, no. 1 (January–February 1988): 103–06.

19. C. Ford Runge, "Positive Incentives for Pollution Control in North Carolina: A Policy Analysis," in *Making Pollution Prevention Pay: Ecology with Economy as Policy,* ed. D. Huisingh and V. Bailey (New York: Pergamon Press, 1982).

f o u r t e e n

Energy Prices and Environmental Costs

Todd Strauss and John A. Urquhart

It has been a direct and troublesome connection. Economic growth is driven by industrial development. Industrial development increases the production and consumption of energy. Energy production and consumption almost invariably lead to environmental damage. We mine or pump coal, oil, and gas out of the ground, altering landscapes and ecosystems. We ship fuels by tanker or pipeline, risking damaging leaks and spills. We burn coal and oil, releasing the pollutants that cause urban smog and acid rain. We build nuclear power plants to generate electricity, creating hazardous waste that we don't know where to store safely. This discouraging picture is made worse by the realization that large parts of the world are only now industrializing and thus are rapidly increasing their consumption of energy.

There are, however, some bright spots. Industry has developed new equipment, new technologies, and new management systems that get more electricity out of each cubic foot of natural gas, more steel using the same ton of coal. Energy intensity, the amount of energy used per dollar of gross domestic product, is decreasing in the industrialized world. At the same time and for many of the same reasons, energy intensity in countries now industrializing is lower than it was for the United States when we were at a comparable stage of economic development.[1]

The energy business has become increasingly integrated, more global, and less regulated. Oil prices are set in commodities exchanges, not in cartel meetings. Price controls have been removed from domestic oil and gas products. The Federal Energy Regulatory Commission (FERC) has opened interstate natural gas transmission and wholesale electricity transmission to competitive market forces. Local electric and gas utilities are merging and transforming into national or even global energy suppliers.

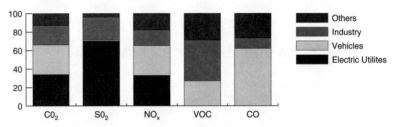

14.1 Sources of U. S. Air Emissions (percentage by pollutant)

From an environmental perspective, these trends present challenges and opportunities. Markets are more efficient when they include the full price of environmental costs and benefits. Although energy deregulation may sharpen market forces, the promise of efficiency depends on "externalities"— such as pollution spillovers—being "internalized," or paid in full. Whether energy deregulation includes appropriate pricing of environmental harms is therefore a critical question.

Although process efficiency has improved and energy markets flourish, the world still has—and will continue to have—petroleum refineries, electric power plants, steel mills, industrial boilers, and automobiles. They discharge carbon dioxide (the major greenhouse gas, associated with global climate change), sulfur dioxide (the major precursor to acid rain), and oxides of nitrogen and volatile organic compounds (precursors to smog). Thus, our attention is focused on air emissions (see fig. 14.1).[2] Inescapably, energy use decisions are environmental policy decisions.

Despite the unbreakable link between the consumption of energy and environmental damage, U.S. energy policy for the last quarter-century has been largely driven not by concerns about the environment but by worries about energy shortages or the national security implications of U.S. dependence on foreign sources of petroleum. Sometimes the responses to these concerns were also environmentally beneficial. For a brief period from 1973 to 1985, high oil prices altered consumption patterns, encouraged energy-efficiency, and stimulated research and development in alternate forms of energy—especially from sun and wind—that cause little environmental damage. Yet this was also a period of "stagflation," with diminished economic growth and high inflation. The challenge now is to have both prosperity and environmental quality.

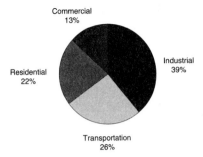

Commercial
13%

Residential
22%

Industrial
39%

Transportation
26%

14.2 U. S. Energy Uses

Pollution from the production and use of energy was not completely ignored, of course. In a society increasingly concerned about how it was going to leave the planet to succeeding generations, the prominent smokestacks of major industrial polluters—"stationary sources" in the regulatory jargon, "big dirties" in the popular lexicon—clearly required regulatory attention. The industrial sector is still the heaviest user of electricity in the United States (see fig. 14.2).[3] Additionally, it was politically expedient to focus more on the highly visible impacts of factories than on the incremental air pollution created by retail consumers (and voters). When air pollution from automobiles was addressed, technological fixes were preferred. Automakers were required to put catalytic converters on cars and petroleum refiners were forced to adjust gasoline formulas to reduce lead and other pollutants.

The focus on big industry and technological fixes produced substantial progress. The level of lead compounds in the air went down dramatically. Emissions of volatile organic compounds decreased. Particulate emissions from electric utilities dropped. But further tightening of existing government regulations—more command and control—does not seem likely to help much. We've gotten the large, easy gains. Cranking down on the existing standards to squeeze out the remaining pollutants is unlikely to yield commensurate gains in air quality, and the economic costs may outweigh the environmental benefits. Moreover, tougher technology standards on new facilities and equipment encourage prolonged use of older and dirtier production processes. Our current regulations often stifle innovation (see chapter 9) and the substantial cost burdens create incentives for foot dragging and delay rather than cooperation and compliance. To get significant further improvement in air quality, our policies need to be more innovative and flexible. The big dirties need contin-ued attention, yet we must pay greater attention to a broader set of energy uses and users, including individuals as car drivers.[4]

Above all, we must recognize that many environmental costs of energy use are not included in market prices. The failure to price the environmental effects of energy use properly compromises the efficiency of our market econ-omy and results in environmental degradation. Capturing or internalizing these costs must be the basis of next-generation environmental policies. The urban air pollution associated with traffic congestion is not included in the market price of driving or riding in an automobile. Adverse health effects are not fully included in existing emission regulations for petroleum refineries. Especially missing, and especially difficult to include, are costs associated with long-distance and long-term effects of air pollution. Although there have been recent changes, states regulate SO_2 emissions to comply with federal standards for local air quality, but do not include the effects of interstate SO_2 transport and consequent acid rain on their neighbors. And so far, the ulti-mate cost of interregional NO_x (oxides of nitrogen) transport and the effects of greenhouse gas emissions have been almost completely missing from the costs of using energy.

There are additional market distortions. Around the world, energy pro-duction and use are frequently subsidized, encouraging excess consumption and often increasing environmental damage. In developing nations, energy is often priced below market cost. In western Europe, national coal industries are typically subsidized. In the United States, poor households may receive fuel oil, electricity, and natural gas at below market prices; production of petroleum (and some alternative fuels such as ethanol) receives tax breaks; both automobile use and mass transit are subsidized; and the Price-Anderson Act limits the liability of commercial operators of nuclear power plants. All these policies exist for social and political reasons. Such policies have envi-ronmental consequences, sometimes good but often bad.

Identifying and including environmental costs and benefits in market prices should be an overarching principle for environmental policy in the energy sector. This is consistent with the trend toward energy deregulation, because it builds on existing and expanding energy markets.

It also represents sound public policy more generally. Prices convey information in the marketplace. If we internalize the full cost of energy production and consumption, we send a clear message to consumers about the environmental damages associated with energy use. This creates a basis for a campaign of public education and awareness. Just as important, there are economic advantages if our policies make costs and benefits "transparent," that is, easily identified, calculated, and understood by users. Transparency makes it easier to incorporate environmental impacts into marketplace decisions. Transparency also reduces the costs of making those marketplace decisions. Finally, transparency can be an effective way of building public awareness. (See chapter 6 for a further discussion of transparency, including the pitfalls of more open policy schemes.)

Taxes or government-imposed fees are the simplest, if not the most politically acceptable, way to adjust energy prices to reflect the costs of environmental damage. Such adjustments are not needed everywhere in the energy market, but at critical points where pollution is created. Our rule should be, the polluter pays. Manufacturers, petroleum refiners, electricity producers, automobile drivers—all pollution-creating energy users should be taxed for the air pollution they emit from burning fossil fuels. Most important, they should be taxed relative to the scale of harms they cause. If not, there will be an inappropriate incentive: to use the energy resource that is less taxed rather than the energy resource that is optimal from a societal perspective.

Because almost everything we purchase—from breakfast cereal to clothes to stereos—takes energy to make or deliver, all prices should reflect the cost of the pollution associated with the energy used in their production and transportation. Producers and consumers alike thus would have an incentive to use less of the pollution-causing energy resource. Because adjusting prices to reflect pollution harms is transparent, perilously so for politicians, policymakers should loudly, clearly, and directly identify energy taxes as pollution charges.

Gasoline

The furor in 1996 over federal gasoline taxes and earlier controversy over a proposed Btu tax might seem to suggest that proposals for higher energy taxes are quixotic, if not politically suicidal. However, the original motivation for both taxes was fiscal, not environmental. The public objected because the taxes were viewed as just another federal burden on taxpayers. The levies

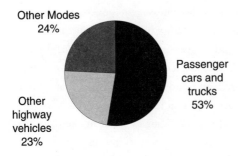

Other Modes
24%

Other
highway
vehicles
23%

Passenger
cars and
trucks
53%

14.3 Transportation Sector Energy Use

were not advanced as a means of reducing environmental damage. The only way to overcome the political hurdles associated with energy taxes is to make the case on environmental grounds. Not only should energy taxes be sold as related to environmental protection; they should also be undergirded with careful science, data, risk assessments, and benefit-cost analyses. With a proper scientific foundation, public support may be easier to develop.

Most energy use for transportation is for passenger vehicles (see fig. 14.3).[5] Thus, energy taxes raise serious equity issues. Higher gasoline taxes, for example, are clearly regressive, since poor people spend a higher percentage of their income on gasoline purchases than wealthy people do. Similarly, gasoline taxes have disparate regional impacts; people in rural areas drive proportionally more than urban dwellers do. These distributional effects must be dealt with if proper energy and environmental policies are to be politically viable. We can consider reducing other taxes on poor people, phasing in pollution charges over time, and adjusting the tax level so that those in rural areas pay for the harms they cause (such as carbon dioxide emissions) but not others (such as some types of smog) that are urban problems.

Voters might also be won over if it were clear that higher pollution taxes would allow other taxes to be lowered. Of course, the public might be skeptical of the political commitment to true revenue neutrality. And some citizens object to energy taxes as just more government intervention in the marketplace. But the logic here is upside down. Not to intervene where this environmental harm exists leaves our economy open to market failure, inefficiency, distorted price signals, and reduced social welfare—in addition to environmental degradation.

Moreover, taxes or other market mechanisms are preferable to command-and-control policies that inflexibly dictate behavior. Consider the recent controversy over reformulated gasoline. Designed to improve urban air quality, reformulated gasoline was required in some parts of the country and optional in other regions. Price increases, purported health and engine problems, and generally unfavorable publicity led to complaints and the withdrawal of some regions from the program. Reformulated gasoline might have been better accepted, and the environmental issue more clearly communicated, if drivers had been offered a choice of reformulated gasoline or standard gasoline bearing a tax of, say, 15 cents a gallon to pay for the harm in another way.[6]

Whether through gasoline taxes or another mechanism, drivers must be held accountable for the air pollution they emit—and the fuel source for 96 percent of transportation energy is petroleum. A per gallon gasoline tax or fee based on miles traveled (an odometer tax) is better than a one-time tax paid by the consumer or manufacturer at the time of the sale. The "use" fees correspond more directly to environmental damage. Driving the vehicle, not owning it, is the polluting activity. (Pollution associated with energy used in manufacturing the vehicle should be reflected in the price of the vehicle.)

Tradable permit programs for NO_x should include motor vehicles. This is a difficult issue, more so politically than technically. Including auto manufacturers, employers, retailers, and local governments in efforts to reduce emissions is sure to lead to creative solutions tailored to particular localities.

Without better public appreciation of energy-related environmental degradation, the demand for cheap gasoline will continue to swamp the desire for environmental protection. The clearest example of political pressure for low-cost energy came in the spring of 1996. Market conditions resulted in temporarily rising gasoline prices. Following public outcry, politicians (from both parties) urged lower gasoline taxes and called for the release of oil from the Strategic Petroleum Reserve.[7] Such reactions were politically expedient but environmentally wrong. Citizens need to know how energy markets work, and need a better understanding of the connection between driving and environmental damage. Without such awareness, there is little hope of a coherent connection between environmental and energy policies.

Electric Power

The U.S. electric power industry—which consumes 36 percent of the fuel used in the United States—presents a similar environmental policy challenge. Burning fossil fuels to produce electricity has significant air emissions (see fig. 14.1). Like gasoline-powered cars and trucks, electric power is a central element of modern society, so any attempt to address environmental harms associated with electric power may have major social, economic, and political consequences.

Until now, electric power has been easier to address politically than driving. Because electric utilities have been regulated monopolies serving franchise territories, government officials have had the means to force electric utilities to include many social and environmental costs in their pricing. Besides "lifeline" prices for the poor and special concessions to attract industry to a region, state regulators (and federal legislators) have been able to encourage or require the use of new technologies and renewable fuels. Some states, for example, have required utilities to offer consumers rebates on energy-efficient appliances.

Now the wave of deregulation that has swept other industries is washing through the power industry. In 1994, the California Public Utilities Commission proposed a competitive market for the state's electricity industry, both at the wholesale generation level and for sales to retail customers. Since then, electricity deregulation, which originally had been driven by large industrial customers clamoring for lower electricity rates, has acquired a momentum of its own as other states have followed California's lead. FERC and some in Congress have also taken active roles in the process. Upheaval is in the works as the industry's longstanding organizational structure is disassembled and realigned. The ostensible goal of this purposeful restructuring is to increase economic efficiency by promoting competition.

Many observers fear that a narrow, pecuniary perspective will prevail, and the environment will suffer in the fray. Public utility commissions are losing some ability to force environmental costs into electric rates. Competition based purely on commodity price will work against new investment in photovoltaics and windfarms, which are costlier than fossil fuel technology but have much less damaging environmental impacts. Continued utility investment in energy conservation and energy efficiency is also threatened.[8]

Wherever the restructuring process leads, two main factors will continue to influence the amount and kind of air emissions from electric power genera-

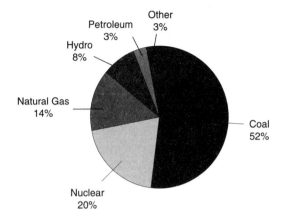

14.4 Fuels for Electricity in the U. S.

tion. The first factor is choice of fuel and technology at individual power plants. The second factor is the total quantity of electricity produced (see fig. 14.4).[9]

First, consider fuels and technology. Natural gas is the cleanest-burning fossil fuel, and is in plentiful supply at reasonable prices. Energy-efficient, combined-cycle gas turbines have become the technology of choice for new power plants. Coal is the dirtiest fossil fuel, but the rates at which SO_2 and NO_x are pushed into the air vary widely among coal-burning power plants. Emission rates depend on the kind of coal used, the way it is burned, and the smokestack cleanup technology employed. Currently, coal supplies over half of the fuel for electricity generation in the United States (see fig. 14.4). To a large extent, the environmental consequences of electricity deregulation depend on how the relative shares of coal and natural gas change.

Environmental policies could shape the outcome. Market-oriented mechanisms could even work in tandem with electricity deregulation, fostering both economic efficiency and environmental protection. Consider the "tradable permit" program for SO_2 emissions created by the 1990 Clean Air Act Amendments. Not only does the amount of SO_2 emitted vary from plant to plant, so too does the cost of controlling these emissions. In a ground-breaking departure from technology-forcing regulation, the tradable permit program allows those plants best positioned to reduce emissions to "overfulfill" their obligations and "sell" the "excess" of their compliance. (See also chapter 7.)

Specifically, the program limits or "caps" the level of total annual SO_2

emissions in the United States. Portions of this total are allocated, according to historical SO_2 emissions levels, to each electric power plant burning fossil fuel. The plants receive their allocated share as allowances, each of which entitles the holder to emit one ton of SO_2. Allowance owners can carry allowances forward from year to year, shift the allowance from one plant to cover emissions from other plants they own, or, most important, buy or sell them. There are heavy penalties for emitting more tons of SO_2 than the allowances one possesses. The owner of a power plant with emissions over its allocation thus has a choice: invest in more air-cleaning equipment, switch fuels, or buy allowances from another plant that has reduced its own emissions below its target level. The seller is therefore rewarded or compensated for its investment in pollution reduction. The buyer is given an economic incentive to invest in new technology, since it will then stop paying for extra allowances and possibly have some excess of its own to sell. SO_2 emissions remain subject to local, state, and federal ambient air quality standards.

The national cap on SO_2 emissions is being lowered in phases to cut annual emissions of SO_2 from power plants to half the 1980 level by 2005. With allowance trading, this environmental target is being attained at lower cost than the utilities would have incurred—and passed on to electricity consumers—if command-and-control regulations had specified a particular pollution abatement technology for all power plant smokestacks. Each utility has an incentive to choose the least costly compliance strategy for its system. Annual savings achieved through this program are estimated at several billion dollars. The program has created competition among various ways to reduce SO_2 emissions, including emission abatement technologies, the use of lower-sulfur coal, and blends of different types of coal. One result: utilities have adopted alternate compliance strategies in greater quantities and more rapidly than expected and, since demand is less, the trading price of an allowance is much lower than originally anticipated.

Whatever the particular details of electricity deregulation, we can be confident the SO_2 cap will be in place. In contrast, imagine that the 1977 Clean Air Act Amendments were still in place. All new coal plants would be required to install expensive smokestack cleanup equipment, but grandfathered plants without the equipment would continue to emit large quantities of SO_2. Under deregulation, such grandfathered plants would then be able to take advantage of their lower costs (from not having to control SO_2 emissions) to sell their electricity in newly competitive markets. The environmental regulation would perversely make the oldest, dirtiest coal plants the fittest to survive and triumph under competition, and the likely result would be increased SO_2 emissions. Under the tradable permit program now in place, such plants are responsible for cutting SO_2 emissions or paying other plants to do so, and no increase in SO_2 emissions can occur.

A guiding environmental principle for electricity deregulation emerges from this case: no electricity producer should be able to obtain a competitive edge by creating pollution—to air, water, land, or habitat—for which it is not held accountable. Currently, power producers pay, at least partially, for local air pollution and for the regional acid rain effects of their SO_2 and NO_x emissions. They are not fully charged for their carbon dioxide emissions, nor for some effects associated with airborne discharge of heavy metals such as mercury.

Besides coal and natural gas, nuclear power will also be greatly affected by shifts to a more competitive electricity market. From an environmental perspective, nuclear power has both advantages and drawbacks. Although nuclear power does not have any CO_2, SO_2, NO_x, VOC, CO, or particulate matter emissions, a nuclear accident may spew radioactive material into the air. Because of this serious but remote possibility, and because of issues associated with disposal of radioactive waste, the environmental profile of nuclear power is substantially different from technologies burning fossil fuel. Should nuclear power have a place in the nation's future energy supply? The answer depends on one's views of the relative risks associated with global warming, acid rain, urban smog, nuclear plant accidents, and radioactive waste disposal. The issue may appear moot, since no additional nuclear plants are on the horizon in the United States. But in the context of electricity deregulation, the relevant environmental question is when currently operating nuclear plants will be decommissioned. The sooner they are taken out of service, the more likely it is that air emissions will increase because the electricity they produce will be supplied from fossil fuel sources instead of renewables or energy-efficiency measures. Early decommissioning would also accelerate the need to manage significant quantities of radioactive wastes. If decommissioning occurred later, there would be more radioactive wastes and increased opportunity for accident, but potential for improved waste handling systems.

As for other technologies and fuels, almost all feasible large hydroelectric sites in the United States have been developed. Solar and wind power currently comprise less than 1 percent of total U.S. electricity supply. They have no polluting air emissions and no fuel costs. Localized placement of photovoltaic panels and wind turbines may be competitive in deregulated U.S. electricity markets. The biggest potential for solar and wind is, however, in developing countries where little or no electricity grid exists, rendering centralized fossil fuel electric generating stations less economical.

A promising, new, clean commercial technology is the fuel cell, which generates electricity by using hydrogen in a chemical reaction, rather than directly burning the fuel. Because fuel cells can deliver neighborhood-sized amounts of electricity and are modular, they have the potential to transform the electric power industry into a decentralized and much less polluting industry.

The second factor determining levels of air emissions is total quantity of consumption. With electricity generated predominantly from fossil fuels, the more electricity we use, the more air pollution is discharged. Deregulation will likely drive prices down, especially for large industrial and commercial customers. Therefore, consumption is likely to rise, with a consequent effect on air emissions.

Similarly, lower electricity prices will reduce consumer interest in energy efficiency. Yet customers care less about the exact quantity of energy they consume or what they pay in cents per kilowatt-hour (kwh) than the total costs of the service they desire—the comfort and convenience electricity provides. We want hot showers and cold soda, not Btu and kwh. The consumer who installs heat pumps, motion-sensitive lighting controls, and timer-controlled thermostats to get better service at lower total cost is incidentally reducing consumption and lowering the impact of electricity use on the environment. And so, as restructuring changes the electric power industry, policymakers should find ways to focus electric utilities on selling services rather than sheer energy. Under the old system of regional monopolies and government-set profit as a return on capital investments, a utility's best interest was to generate and sell as many kilowatt-hours as possible. As this arrangement collapses, the way needs to be cleared for competition between existing electric utilities and new or expanded energy service companies in the provision of systems and equipment that improve consumer service while lowering electricity usage. The policy conclusion: a restructuring model that promotes retail competition is desirable.

Retail competition would foster opportunities for sellers to market "green energy," electricity derived from sources less damaging to the environment, such as wind power or photovoltaics. Green energy is being marketed now around the country, in many of the experimental trials of retail electric competition. The size of the market for green energy is unknown, but the potential is significant. Of the households and companies that now recycle, how many might pay the same or a little extra to power their electrical appliances and machines with green energy? This kind of marketing, sorely lacking when vertically integrated regulated electric monopolies dominated the landscape, may be the best hope for renewable resources in the United States.

Finally, even with a restructured electricity industry, the current consensus is that because of economies of scale, transmission and distribution of electricity—the "wires" business—will continue to be provided by regulated monopolies. And so there will remain opportunities for regulators to insert environmental costs and benefits into the electric power system.

Institutions

Federal and state regulatory commissions charged with the economic regulation of energy need to broaden their viewpoint. In regulating public utilities, these commissions have attempted to fulfill the role of the missing market. As energy markets take shape and mature, regulators need to fulfill the role of a missing market for environmental quality, ensuring that energy suppliers pay for the environmental harms they cause.

To support local or state efforts to capture externalities into prices, and to protect against uncompensated interstate pollution, the Environmental Protection Agency should develop policy tools to properly identify and analyze the effects and costs of pollution on human health and natural ecosystems. Since many important environmental problems have a global dimension, it is especially important that the EPA work with its counterparts in other countries, and with the appropriate international organizations, to reflect the full costs and benefits of policy choices. Currently, an independent Energy Information Administration in the Department of Energy produces high-quality energy market information and analysis. A counterpart bureau should be established in the EPA with a similar mission and analytic focus. Because it would assess, in economic terms, the environmental impacts and risks to human and ecosystem health, the tasks of the bureau would be difficult and controversial. Nevertheless, the federal government clearly needs policy-neutral but policy-relevant environmental analysis, especially if we are to rely more on pricing pollution harms.

Wherever possible, we need to replace the old command-and-control policies with more effective and efficient market-oriented policies. But when this is not practical or politically possible, the old programs should be refined for greater cost-effectiveness, not simply scrapped. The cornerstones of next-generation energy policies should be full-cost energy pricing through environmental taxes, tradable permits, and other market mechanisms. Higher environmental quality and more economically efficient use of energy resources should result. This makes ever greater sense as energy markets become more integrated, more global, and less regulated.

Notes

1. José Goldemberg, "Energy and Environmental Politicies in Developed and Developing Countries," *Energy and the Environment in the Twenty-first Century*, ed. Jefferson W. Tester et al. (Cambridge: MIT Press, 1991).

2. Percentages for CO_2 based on data in U.S. Energy Information Administration, *Emissions of Greenhouse Gases in the United States 1995,* DOE/EIA-0573(95) (Washington, D.C.: Government Printing Office, 1995); percentages for other pollutants based on data in U.S. Environmental Protection Agency, *National Air Pollutant Emission Trends, 1900–1994,* 454-R-95-011 (Washington, D.C.: Government Printing Office, 1995).

3. Calculated from data in U.S. Energy Information Administration, *Annual Energy Review 1995* (Washington, D.C.: Government Printing Office, 1996). U.S. energy consumption for 1995 is about 90 quadrillion (10^{15}) Btu, or about one-quarter of total world energy consumption.

4. As individuals, we make our greatest contribution to pollution every time we pull out of the driveway. See chap. 12 for a discussion of next-generation policies to address transportation.

5. Calculations are based on data in Stacy C. Davis and David N. McFarlin, *Transportation Energy Data Book: Edition 16,* ORNL-6898 (Oak Ridge, Tenn.: Oak Ridge National Laboratory, 1996).

6. In "A Cheaper Way to Clean Gasoline" (*POWER Notes,* University of California Energy Institute [Fall 1996]), Severin Borenstein and Steven Stoft argue for such a tax for a different reason—to mitigate price spikes in the California gasoline market.

7. Leon Jaroff et al., "Fuming over Gas Prices; Politicians Are Jumping to Intervene, But High Prices May Be the Result of Increasing Demand," *Time,* 13 May 1996.

8. "Moody's Downgrades EUA Cogenex Notes; Cites Continuing Decline in Utility DSM," *Energy Services and Telecom Report,* 24 Oct. 1996.

9. U.S. Energy Information Administration, *Annual Energy Review 1995* (Washington, D.C.: Government Printing Office, 1996), table 8.2.

fi f t e e n

A Vision for the Future

Daniel C. Esty and Marian R. Chertow

Thinking Ecologically offers a set of ideas on which a new American environmental policy structure might be built. Wide-ranging and not entirely consistent in the course they set, the thoughts and concepts introduced in the preceding chapters are meant to stimulate discussion rather than lay out a specific action plan. In fact, we fully expect that although some of the ideas may find their way into next year's environmental debates, others may take decades to percolate through the policy-making process.

We recognize, moreover, that there are potential contradictions between and among some of the key ideas. For example, can Carol Rose's call for greater emphasis on environmental property rights be squared with the more community-minded approach to managing resources advanced by John Turner and Jason Rylander? We think so. We see a role for property rights in helping to define the bounds of acceptable environmental behavior in a manner that is not inconsistent with an emphasis on greater cooperation in pursuit of collective goals. Rose's focus on public as well as private property rights makes this compatibility even clearer. More broadly, we envision a policy regime that draws on a wide range of strategies and approaches, reflects the spectrum of issues that must be dealt with under the environmental rubric, and applies different tools and specific policy mechanisms depending on the problem at hand.

We also realize that we have not answered all of the hard questions that face next-generation environmental policymakers. Emil Frankel's call for cars and transportation systems that are less polluting does not address the habitat destruction of road systems, which is a central focus of the land use chapter and an implicit concern in John Gordon and Jane

Coppock's discussion of ecosystem management. We see an opportunity to create incentives for thinking which will, over time, link transportation and development decisionmaking. With a more comprehensive understanding of their job and a better appreciation of how their work fits into a broader set of societal and natural systems, highway planners could come to see issues such as habitat destruction as design considerations.

The obstacles to achieving consensus on a new policy course cannot be minimized. Finding the political will for change is always difficult. A set of firmly entrenched interests supports the status quo. Americans are deeply imbued with private-property and individual-rights notions that can seem irreconcilable with the collective action that is often required to advance common-good solutions to environmental problems.

The move to more effective and efficient policies is made especially arduous by a set of equity issues that often are embedded in the current regulatory structure. Hiking gasoline taxes, for example, raises hackles because low prices operate as a subsidy—a particularly important one for poor people (as well as inhabitants of rural areas and others who drive a lot). In a country that prizes the image of self-sufficiency and generally looks askance at programs that redistribute income, this hidden "support" goes unquestioned. Raising energy taxes and thus the price of gasoline (without attention to distributional effects) threatens to leave significant portions of the population with less money in their pockets. No matter how significant the environmental gains to be achieved with energy charges that reflect pollution costs, such policies will face tough political sledding until the objections of those who perceive themselves to be losers from higher gasoline prices are taken into account. The similar equity claims of farmers, small business people, property owners, and all those whose economic circumstances and sense of entitlement would be frustrated by new approaches to environmental protection must also be addressed if next-generation policies are to advance.

In the past, when environmental insults were obvious and the targets of controls were big smokestack industries, making companies pay for their despoliation had a moral logic that offered wide appeal. Today, however, when many of the harms we face reflect the cumulative impact of millions of individuals and small enterprises, the enemy is "us" and the moral certainty of a crusade is harder to sustain. The politics of change are never easy.[1] But fashioning a coalition to carry forward environmental policy will be especially challenging.

More supple policy approaches that can be tailored to varying circumstances may help to make new and refined environmental programs possible. A more *ecological* approach to policymaking that recognizes the connections that must be made from environmental goals to other aspects of life should alleviate some of the political tensions that have weighed down recent environmental reform initiatives. But disputes over values are inevitable, and the accompanying political intrigue will persist. In fact, Robert Socolow, looking forward a century, suggests that the environmental future "will be a no less restless time than our own."[2] He is undoubtedly correct. We can anticipate different problems, but we should not expect fewer. Nevertheless, understanding the world as a complex series of interconnected systems provides a perspective that can make the reconciliation of competing interests and conflicting desires more achievable.

Broadening the scope of policy-relevant considerations can also help to blunt the current tendency to overlook the secondary impacts or countervailing effects of our regulatory interventions. But a more comprehensive policy approach will not eliminate boundary problems. Line drawing will still be necessary, and some harms and effects will still have to be deemed too indirect or remote to factor into the policy calculus.

It is important, of course, that reforms promising greater flexibility in support of beneficial innovation not become an excuse for shifting pollution costs onto the public or for inattention to the risk of environmental disasters. As always, the devil lies in the details of the proposals that go forward. Market forces, for example, offer the prospect of improved environmental performance at lower costs—but market mechanisms can also become a mere slogan parroted by special interests seeking preferred treatment in the regulatory process. With these pitfalls in mind, it becomes clear that laying the analytic foundations for a next generation of environmental policy will take careful and systematic attention.

How would the world look if policies incorporating the ideas of *Thinking Ecologically* were implemented? At the risk of appearing utopian but, in fact, cognizant of the complexity of the environmental reform challenge, we offer the following scenario of the future, circa 2020. Our vision is deliberately optimistic, and by no means the most likely. But we think it is useful as an illustration of the changes in day-to-day life that might unfold with the right mix of next-generation policies:

- Environmental protection has become everybody's business. Under a commitment to inclusive environmental decisionmaking, federal, state, and local officials work with industrial organizations, environmental groups, businesses, and civic associations to make policy choices match community needs and conditions. From city managers to highway planners to retailers, public and private sector officials understand their position as environmental decisionmakers and take seriously their stewardship role. Many citizens are involved in public advisory groups empowered to guide the environmental management of nearby rivers, lakes, bays, open space, and transportation corridors.

- Environmental policy has become more ecological—comprehensive in focus and attentive to linkages across problems. Systems thinking in the form of both industrial ecology and ecosystem management has emerged as the analytic core of ecological policy. Fragmented regulatory approaches derived from individual laws separately governing air and water pollution and waste management have, over time, been reassembled, omitting some parts, adding others, into a more coherent and unified set of obligations. Similarly, a focus on the integration of human and natural systems, with its broad perspective on the competing needs and desires of the public, gives concreteness to commitments to sustainable development.

- In the land use context, this new policy approach provides a mechanism to address the cumulative impacts of many small harms and thus to ensure that environmental goals are better connected to development decisions. With an emphasis on comprehensive analysis and data-driven decisionmaking, new procedures gauge the air, water, and habitat impacts of proposed land uses. This process supports local priority setting and helps to guarantee that any environmental burdens created are not unfairly imposed on those in the next town, in the next state, or even thousands of miles away.

- Market economies thrive around the world and increased wealth has made many countries more attentive to environmental concerns. At the same time, the public in the United States and elsewhere recognizes that market forces only yield efficient results if a system of checks on pollution by companies and individuals is in place. Thus, while the accepted domain of government activity has shrunk, there is widespread appreciation that environmental protection stands alongside national security as a core governmental function.

- While responsibility for some policy matters has been decentralized, the "spirit of regulatory devolution" from the 1990s has given way to a recognition that the diversity of environmental and resource use issues requires a diversity of responses. Some problems are known to be best dealt with at a community or company level; other issues require a national or even global response. Local, state (and Indian tribe), and federal activities are supplemented by multitown, multistate, and multination compacts where the ecosystem scale dictates.

- Environmental rights—the entitlement of every person to be free from pollution harms—have been firmly established, actually reestablished, building on the tradition of nuisance law. The property rights of the public have similarly been clarified, and landowners recognize that they must pay for harms that spill beyond their property boundaries or for any scarce common resources such as air or water that they consume or pollute.

- A carefully structured system of fees for emissions has been established. Payment is required for discharges that cause adverse ecological or public health effects above established thresholds. After a phase-in period that extended twenty years in some sectors, companies from multinational giants to mom-and-pop enterprises pay emissions fees. Similarly, users of public resources such as water and grazing land are charged market prices for their consumption. Even farmers have gotten used to paying full price for their water and "nitrogen-loadings" fees on their fertilizer purchases.

- Individuals, as well, pay for their pollution. Pay-as-you-throw garbage charges have become a universal standard, and the prospect of a flat fee for trash pick-up seems as humorously outdated as unmetered electricity might have in the 1990s. Other "green" fees have also become commonplace: "highway bills" for those commuting on major roads (many of which have now been privatized—improving maintenance and decreasing congestion); electronically collected smog tolls for those driving polluting vehicles in areas with substandard air; and greenhouse gas taxes on gasoline, fuel oil, and natural gas (scaled to reflect the relative climate change impacts of the various fossil fuels).

- Tax-free electric cars have become popular among commuters and most fleet vehicles run on natural gas to reduce the environmental fees they must pay. General Motors, spurred by the changing economics of engine fuels, has a full line of hydrogen-powered vehicles.
- The income tax has been abolished. A consumption tax—with rates determined by annual family purchases—augmented by revenues from pollution fees has replaced it. The new revenue structure encourages savings and investment and simultaneously promotes resource conservation. Equity concerns over the regressive nature of pollution taxes and the fear that poor people will not be able to pay fees such as highway use charges have been addressed through a graduated consumption tax schedule under which the first twenty thousand dollars in purchases for an average household is tax-free.
- Systems thinking has led to a reorganization of federal policymaking. The former Environmental Protection Agency, Department of Energy, Forest Service, Interior Department, National Oceanic and Atmospheric Administration, Food and Drug Administration, and Occupational Safety and Health Administration have been consolidated into a new Public Health, Environment, and Resources Department (PHER, or "Fair," as it is called). PHER's central role is evaluating and putting a price on public health and ecological harms.
- PHER's work is supported by a Bureau of Environmental Indicators and Statistics and a reconfigured independent scientific agency, the National Institutes of Health and Environment (NIHE). The NIHE scientists, drawn from a variety of disciplines, have garnered a reputation for anticipating new environmental and health issues, and even a grudging respect from regulated industries for the rigor and transparency of their technical analyses.
- Environmental officials at all levels place policy emphasis on delivering maximum environmental value for the resources, both public and private, devoted to environmental protection. A variety of mechanisms (such as multiyear phase-ins for new regulations, or transitional "adjustment support" funds) make the regulatory structure effective yet flexible.
- Although the enforcement officials still must pursue a small set of industry laggards, most companies operate under a command-and-

covenant system through which their compliance with established environmental performance goals is monitored by independent auditors. Deficiencies are reported to the government and result in enforcement actions.

- In many cases, environmental protection efforts proceed without government involvement. Companies pay close attention to their "resource productivity" and competitive advantages gained through technological innovation. Corporate cost accounting practices have been refined to permit close tracking of all materials and energy flows. The International Organization for Standardization's environmental management guidelines (ISO 14000) from the 1990s have flowered into the ISO 28000 series of substantive product and production process requirements.

- Moving beyond the end-of-pipe pollution controls of the 1970s and even the pollution prevention efforts of the 1990s, businesses now analyze comprehensively the environmental harms and risks created by all of their facilities and throughout the life-cycles of their products. Sophisticated computer models linked to vast impact databases give a methodological rigor to life-cycle analysis that seemed unimaginable in the late twentieth century. Moreover, the systems focus encouraged by industrial ecology–based policies promotes careful corporate consideration of opportunities to reduce environmental impacts across the spectrum of activities of the firm, as well as upstream (with suppliers) and downstream (with customers).

- Many parts of the business community stand at the leading edge of ecological and public health advances. Indeed, in the developing world, countries vie for the investment of multinational companies known to bring advanced pollution prevention and control technologies, environmental management systems, and environmental education programs with them. Private capital flows are recognized to be the main engine of sustainable development in both the industrialized and developing worlds. Environmental provisions (multitiered performance standards set to reflect differences in development and environmental circumstances) have been folded into the Multilateral Agreement on Investment (MAI) to ensure that countries do not compete for private capital on the basis of environmental degradation.

- The "customer is always right" motto has been given new meaning with the introduction of an "eco-facts" label on most consumer products. Shoppers compare product pollution profiles just as their relatives in the late 1990s contrasted fat contents on nutrition labels among competing supermarket brands. The Webzine *EnviroReports* competes with *Consumer Reports* as a source of more detailed information on environmental impacts and product performance. Internet information on how to make purchases in the most eco-sensitive way is also available. Should an environmentally conscious buyer shop at a regional mall, visit a local store, or purchase from the home shopping network? Ask the computer.

- Many corporations have adopted voluntary "good neighbor" environmental programs that include continuous emissions monitoring, regular environmental outreach meetings with community leaders and the general public, and the posting of minute-by-minute emissions release data on the Internet as well as on electronic scoreboards at the factory gate.

- The government too has adopted a good neighbor policy. Federal land managers give careful attention to the needs of communities on or near federally owned properties. Moreover, a federal land sales program in the West (responding to public distress over the large percentage of government-owned land) has generated funds for land acquisition in the East, where the opposite sentiment (that too little land is government-protected) exists.

- The Harvard Business School case study on a late twentieth-century environmental policy tangle called "Superfund" highlights the danger of misplaced regulatory responsibilities, too narrow vision, and the failure to take economic context into account. Many older factory sites, abandoned in the 1980s and 1990s because of fears of liability for past contamination, have now been revitalized with community-determined clean-up requirements that reflect future land use plans. In fact, bankers now frown on "greenfields" proposals for industrial expansion, recognizing the cost burden for new water, sewer, and road systems as compared to reusing existing infrastructure.

- The marriage of information technology and environmental protection has begun to bear fruit. Developers, for example, can go into any land records office in the country and "click" on computer maps of parcels in which they have a potential interest. Drawing on land records, GIS mapping, and other data inputs, the screen reveals information on past uses of the property, including possible contamination, geological site limitations, and potential wetlands and endangered species issues. This database allows potential property purchasers to make informed decisions and permits greater predictability and speed on the part of government regulators.

- If a proposed building site contains a wetland or other ecologically important area, the developer instantly learns its "value rank." Although permission to build on red maple swamps or other high-value habitats is severely restricted, construction is permitted on low-value sites after a contribution (commensurate with the value of the habitat loss) is made to a wetlands or "ecosystem conservation" bank. The bank uses the funds to purchase high-value properties for protection and also to restore damaged wetlands as an offset against the habitat loss from new construction.

- Internationally, a half-dozen U.N. agencies and various treaty secretariats with environmental missions have been consolidated into a Global Environmental Organization (GEO). This new, leanly staffed body manages the international response to global-scale problems, provides a mechanism for data and information exchange, coordinates policy with other international bodies such as the World Trade Organization (WTO), and offers dispute settlement services for transboundary environmental issues. GLOBE, a group of environmentally interested legislators from around the world, conducts oversight hearings on the GEO's performance.

- The World Bank and other multilateral development banks have been rechartered. They now focus on assistance to the least developed countries (those unable to attract private capital) and on subsidizing the global benefits of environmental investments in projects where a nation-state level benefit-cost analysis would not justify action because the benefits fall outside the country.

- Environmental nongovernmental organizations (NGOs) help to shape and legitimize international environmental policymaking. As part of a rapidly evolving global civil society, the NGOs link local citizens to the otherwise distant governance processes at the world scale, inform the public about policy choices, and ensure a measure of public participation in various international organizations such as the GEO and the WTO.

- Everyday citizens are both more aware and less aware of their roles as environmental actors. They recycle as enthusiastically as ever, drop food wastes down the compost chute now built into every modern kitchen, and (in drought-plagued areas) have learned to redirect the gray water from their sinks and showers through a filter and onto their lawns (which are generally smaller than in the twentieth century, to reduce mowing time as well as water use). More dramatically, as prices have begun to reflect the full costs of ecological and public health harms, the invisible green hand of market forces has steered the public toward more environmentally benign products.

- Where once slogans such as "every day is Earth Day" were needed to remind people about caring for the planet, environmental education, both formal and informal, has transformed attitudes. Parents instruct their children about good environmental practices and habits just as children in the late twentieth century educated their parents. There can be no doubt, the next generation has arrived.

Notes

1. Interestingly, "public choice" theorists (e.g., James M. Buchanan and Robert D. Tollison, eds., *The Theory of Public Choice* [Ann Arbor: University of Michigan Press, 1984]) and some analysts of regulatory politics (e.g., James Q. Wilson, ed., *The Politics of Regulation* [New York: Basic Books, 1980]) argue that it should be easier to impose costs on a large and relatively unorganized class of actors, like small business or the general public, who are less likley to be politically mobilized in opposition than big industry. But the costs of pollution prevention and control may not be so small as to be inconsequential, or they may be so conspicuous, regardless of their actual size, that even a diffuse group of putatively hard-to-organize cost-bearers will become politically active.

2. See Robert Socolow et al., *Industrial Ecology and Global Change* (Cambridge: Cambridge University Press, 1994), 16.

Next Generation Project Participants

The individuals listed below contributed to Environmental Reform: The Next Generation Project at the Yale Center for Environmental Law and Policy by participating in the expert workshops and/or assisting the authors in reviewing chapter drafts. Although the names of the organizations with which participants were affiliated at the time of their contribution are listed, each person acted individually and the names of the organizations are given for identification purposes only.

Fred Abbott (Chicago-Kent School of Law)
James T. Addis (Wisconsin Department of Natural Resources)
Richard N. L. Andrews (University of North Carolina, Chapel Hill)
Willis Anthony (Minnesota Corn Research and Promotion Council)
Jim Arts (Wisconsin Federation of Cooperatives)
Nicholas J. Ashford (Massachusetts Institute of Technology, Sloan School of
    Management)
Robert U. Ayres (The European Institute of Business Administration [INSEAD])
Donald J. Barry (U.S. Department of the Interior)
Sandra S. Batie (Michigan State University)
S. William Becker (State and Territorial Air Pollution Program
    Administrators/Association of Local Air Pollution Control Officials
    [STAPPA/ALAPCO])
Vicki L. Been (Harvard University Law School)
Frances Beinecke (Natural Resources Defense Council)
Julie Belaga (Export-Import Bank of the United States)
Jeffrey R. Berry (Aetna)
Anthony J. Drexel Biddle III (Chase Manhattan Bank)
John B. Blatz (Dexter Corporation)
Larry A. Boggs (General Electric Company)
Fred Bosselman (Chicago-Kent Law School)
Lindsey Brace (Hancock Timber Resource Group)
Lee Breckenridge (Northwestern University School of Law)
Esteban Brenes (Instituto CentroAmericano de Administración de Empresas
    [INCAE])

Russell L. Brenneman (Connecticut Forest and Park Association)
Deborah Brown (U.S. Environmental Protection Agency—New England)
Howard Brown (RPM Systems)
Mike Brown (Patagonia, Inc.)
F. Scott Bush (National Environmental Policy Instititue)
J. Peter Byrne (Georgetown University Law Center)
James Cabot (U.S. Environmental Protection Agency–New England)
J. H. Caldwell (Center for Energy Efficiency and Renewable Technologies)
Keane Callahan (Robinson and Cole)
Laura Campbell (Environmental Law Institute)
Jonathan Z. Cannon (U.S. Environmental Protection Agency)
Leslie Carothers (United Technologies)
Armando J. Carbonell (Cape Cod Commission)
Margaret Carson (Enron Corporation)
Luis Castelli (Fundación Ambiente y Recursos Naturales)
Garciela Chichilnisky (Columbia University)
Richard Chow (EnergyWorks)
Timothy W. Clark (Yale School of Forestry and Environmental Studies)
Don R. Clay (Don Clay Associates)
Charles H. Collins (Forestland Group)
Jane Coppock (Yale School of Forestry and Environmental Studies)
Stephen P. Crosby (Smart Route Systems)
Frank L. Danchetz (Georgia Department of Transportation)
Richard A. Dennison (Environmental Defense Fund)
John DeVillars (U.S. Environmental Protection Agency–New England)
Brad DeVries (Land Stewardship Project)
Hank Dittmar (Surface Transportation Policy Project)
Jeff Dlott (Dlott Consulting)
Donna M. Downing (Apogee Research, Inc.)
Brendan Doyle (U.S. Environmental Protection Agency)
Richard Draeger (Pacific Gas and Electric Company)
Dean Drake (General Motors)
E. Linn Draper, Jr. (American Electric Power)
Jeffrey Dunoff (Temple University)
Donald N. Duvick (Iowa State University)
George C. Eads (Charles River Associates)
Ralph Earle (Alliance for Environmental Innovation)
John Echeverria (National Audubon Society)
Robert F. Ehrhardt (General Electric Company)
Robert Ellickson (Yale Law School)
William Ellis (Yale School of Forestry and Environmental Studies)

Jürgen Ertel (Brandenburg Technical University of Cottbus)
David Ervin (Henry A. Wallace Institute for Alternative Agriculture)
Brock Evans (National Audubon Society)
Paul Faeth (World Resources Institute)
Scott A. Fenn (Investor Responsibility Research Center)
Christina Figueres (Center for Sustainable Development in the Americas)
Adam Finkel (U.S. Occupational Safety and Health Administration)
Massimo Fioruzzi (GreenValue)
William A. Fischel (Dartmouth College)
Allen Fitzsimmons (Balanced Resource Solutions)
Robert W. Frantz (General Electric Company)
Hillary French (Fletcher School of Law and Diplomacy, Tufts University)
Eric T. Freyfogle (University of Illinois College of Law)
Robert Friedman (H. John Heinz III Center for Science, Economics, and the
    Environment)
Robert Frosch (Harvard University, John F. Kennedy School of Government)
John Fryderlund (Heritage Foundation)
John T. Ganzi (Environment and Finance Enterprise)
Gregory E. Gardiner (Yale University)
John Gault (Gault and Associates)
Bradford Gentry (Yale School of Forestry and Environmental Studies)
Michael Gerrard (Arnold and Porter)
Kenneth R. Goldberg (Florida Department of Community Affairs)
Deborah Gordon (transportation consultant)
Jane Hotchkiss Gordy (Conservation Law Foundation)
Michael Gough (Cato Institute)
Thomas E. Graedel (Yale School of Forestry and Environmental Studies)
Thomas J. Graff (Environmental Defense Fund)
Elisa Graffy (U.S. Geological Survey Center)
John Graham (Harvard School of Public Health)
George Gray (Harvard Center for Risk Analysis)
Lakshman Guruswamy (University of Tulsa College of Law)
Bart Hague (consultant)
Robert M. Hallman (Cahill Gordon and Reindel)
Louise Halper (Washington and Lee University)
Gerald Hapka (DuPont)
Stuart L. Hart (University of Michigan)
Scott B. Harvey (Association of American Railroads)
George R. Heaton (Worcester Polytechnic Institute)
Miriam Heller (University of Houston)
Ben G. Henneke (Clean Air Action Corporation)

Peter Hennicke (Wuppertal Institut)

Ruth Hennig (John Merck Fund)

Jeremy Hockenstein (McKinsey and Company)

Thomas A. Horan (Claremont Graduate School)

Donald Hornstein (University of North Carolina School of Law)

Lori Hotz (Dexter Corporation)

Walter S. Howes (EBI Capital Group)

Arnold Howitt (Harvard University, John F. Kennedy School of Government)

James L. Huffman (Northwestern University School of Law)

Frederick Huntsberry (General Electric Company)

Bryan Husted (Instituto Tecnologico y de Estudios Superiores de Monterrey)

Barry Ilberman (Northeast Utilities)

Leo Jensen (Dutch Sustainable Technology Programme)

Frank Joyce (Ecotech Research and Consulting, Ltd.)

Myron Just (agriculture consultant)

Hal Kassoff (Maryland Department of Transportation, State Highway Administration)

Robert Keiter (University of Utah College of Law)

Dorothy A. Kelly (Ciba-Geigy Corporation)

Robert C. Kelly (Amoco/Enron Solar Power Development)

Robert R. Kiley (New York City Partnership and Chamber of Commerce)

Judith B. Krauss (Yale School of Nursing)

Judith M. LaBelle (Countryside Institute at Glynwood Center)

Thomas D. Larson (consultant)

Howard A. Latin (Rutgers State University)

Richard Lazarus (Washington University)

Douglass Lea (consultant)

James Lee (American University, School of International Service)

Reid J. Lifset (Yale School of Forestry and Environmental Studies)

K. R. Locklin (E and Co.)

Raymond B. Ludwiczewski (Gibson, Dunn and Crutcher)

Joan Licht Mantel (Global Project and Structured Finance Corporation)

Theodore T. Marmor (Yale School of Management)

James M. McElfish, Jr. (Environmental Law Institute)

G. Tracy Mehan III (Michigan Department of Environmental Quality)

Robert Meltz (Library of Congress)

Dwight H. Merriam (Robinson and Cole)

Dini S. Merz (Emily Hall Tremaine Foundation)

Stephen M. Merz (Yale–New Haven Hospital)

Granger Morgan (Carnegie Mellon University)

Dick Morgenstern (Resources for the Future)

Jeffrey C. Muffat (3M Environmental Technology and Services)
David T. Musselman (Cinergy Corporation)
William G. Myers III (National Cattlemen's Beef Association)
Roger Naill (AES Corporation)
Linda Descano Nelson (Salomon Inc.)
Klaus W. Nielsen (United Parcel Service)
David Olsen (Center for Energy Efficiency and Renewable Technologies)
Hari Osofsky (Global Environment and Trade Study)
David E. Osterberg (Cornell College)
Robert Paarlberg (Wellesley College)
Timothy J. Penny (University of Minnesota)
Robert Percival (University of Maryland, Baltimore)
Rutherford H. Platt (University of Massachusetts, Amherst)
Nigel Preston (North West Water International Limited)
Carlos E. Quintela (Avina Foundation)
Alan Charles Raul (Beveridge and Diamond)
David W. Rejeski (Executive Office of the President, U.S. Office of Science and
    Technology Policy)
Renate Rennie (Tinker Foundation)
Michael A. Replogle (Environmental Defense Fund)
Alison Rieser (University of Maine School of Law)
Chris Risbrudt (USDA Forest Service)
Mark Ritchie (American Farm Bureau Federation)
Jorge Rivera (Instituto CentroAmericano de Administración de Empresas [INCAE])
David B. Rivkin, Jr. (Hunton and Williams)
Andrea Silvana Rodriguez (London School of Economics)
David Roe (Environmental Defense Fund)
G. Jon Roush (consultant)
David S. Rubenson (Rand Corporation)
Michael Rubino (International Finance Corporation)
James W. Rue (Pennsylvania Department of Environmental Protection)
Frances M. Rundlett (Potomac Hudson Engineering, Inc.)
David B. Sandalow (White House, Council on Environmental Quality)
Michael T. Saunders (U.S. Department of Transportation, Federal Railroad
    Administration)
Hank Schilling (General Electric Capital)
Joseph Schilling (National Environmental Policy Institute)
Frederick Schmidt-Bleek (Wuppertal Institut)
Camilla Seth (Environmental Advantage)
Nicholas Shufro (United Technologies)
David D. Sigman (Exxon Chemical Company)

Jody Sindelar (Yale School of Epidemiology and Public Health)

David Sive (Sive, Paget and Riesel, P.C.)

Edmund J. Skernolis (WMX Technologies, Inc.)

Robert E. Skinner, Jr. (National Research Council)

Lisa Skumatz (Skumatz Economic Research Associates)

Wick Sloane (consultant)

William R. Slye (Pace University)

Robert H. Socolow (Princeton University)

William Stigliani (University of Northern Iowa)

William Stillinger (Northeast Utilities)

Bruce N. Stram (Enron Capital and Trade Resources)

Maurice F. Strong (United Nations)

Arthur J. Swersey (Yale School of Management)

Dag Syrrist (Technology Funding)

Pedro Tarak (Fundación Ambiente y Recursos Naturales)

A. Dan Tarlock (Queensland University of Technology)

Edward Thompson, Jr. (American Farmland Trust)

Linda Thrasher (Cargill, Inc.)

Lloyd Timberlake (Avina Foundation)

Bruce D. Tobecksen (WMX Technologies, Inc.)

Hoo-min Toong (Quantum Energy)

Héctor R. Torres (Permanent Mission of Argentina to the World Trade
    Organization)

E. Thomas Tuchmann (U.S. Department of Agriculture)

Judy Tykociski (McKinsey and Company)

Laura Underkuffler (Duke University School of Law)

Joanna D. Underwood (Inform, Inc.)

Charlotte Uram (Sidley and Austin)

David Vadon (Salomon, Inc.)

Ariane van Buren (Interfaith Center on Corporate Responsibility)

Macelo Drügg Barreto Vianna (Instituto Cultural e Filantrópico Alcoa)

David Vogel (University of California, Berkeley, Haas School of Business)

Konrad von Moltke (Dartmouth College)

Deborah Vorhies (U. N. Environment Programme)

William Vorley (Leopold Center for Sustainable Development)

Sarah Wade (Hagler Bailly)

Karl Wagener (Connecticut Council on Environmental Quality)

Kathryn D. Wagner (University of Pennsylvania)

Thomas W. Wahman (Resources Development Foundation)

Bill Walsh-Rogalski (U.S. Environmental Protection Agency–New England)

John P. Wargo (Yale School of Forestry and Environmental Studies)

Tim Warman (American Farmland Trust)

Robert F. Weltzien (Roots, Inc.)

Jake Werksman (Foundation for International Environmental Law and Development, London)

Mark A. White (University of Virginia, McIntire School of Commerce)

Mason Willrich (EnergyWorks)

Jim Wilson (Resources for the Future)

Sebastian Winkler (U.N. Environment Programme)

Lawrence Yermack (Parsons, Brinkerhoff)

Richard A. Zais, Jr. (City of Yakima, Washington)

Elaine Y. Zielinski (U.S. Department of the Interior)

Jeffrey M. Zupan (Regional Plan Association)

appendix two

Contributors

Coeditors

Marian R. Chertow is director of Environmental Reform: The Next Generation Project and has been on the faculty of the Yale School of Forestry and Environmental Studies since 1990. She worked in state and local government for seven years, including service as president of the Connecticut Resources Recovery Authority. Her policy expertise is in waste management, technology innovation, and business/environment issues.

Daniel C. Esty is director of the Yale Center for Environmental Law and Policy. He is an expert in trade and the environment and was at the U.S. Environmental Protection Agency from 1989 to 1993, including service as deputy assistant administrator for policy, planning, and evaluation. Professor Esty has a joint appointment at the Yale Law School and the Yale School of Forestry and Environmental Studies.

Authors

Steve Charnovitz is director of Global Environment and Trade Study, a policy consortium located at Yale University. He was previously policy director of the U.S. Competitiveness Policy Council and legislative assistant to the Speaker of the U.S. House of Representatives.

Jared L. Cohon is the president of Carnegie Mellon University. He was the eighth dean of the School of Forestry and Environmental Studies and professor of environmental systems analysis at Yale University. His research and teaching focus on the development of systems analysis techniques and their application to the management of environmental and resource problems.

Jane Coppock is assistant dean of the Yale School of Forestry and Environmental Studies. She has conducted projects with the Nature Conservancy and the United Nations Development Programme. She was formerly on the humanities faculties of Dartmouth and MIT.

Elizabeth Dowdeswell is the executive director of the United Nations Environmental Programme, headquartered in Nairobi, Kenya. Before joining the United Nations, she was assistant deputy minister at Environment Canada and head of the Atmospheric Environment Service.

E. Donald Elliott is senior partner at the law firm of Paul, Hastings, Janof-sky, and Walker. A member of the Yale Law School faculty since 1981, he also served as assistant administrator and general counsel of the U.S. Environmental Protection Agency from 1989 to 1991.

Emil Frankel is of counsel to the law firm of Day, Berry, and Howard. He specializes in issues of transportation, environmental policy, and privatization of transportation facilities. He was commissioner of the Connecticut Department of Transportation from February 1991 until January 1995.

John C. Gordon is Pinchot Professor of Forestry and Environmental Stud-ies at Yale. His current research is on environmental change and biological diversity in Alaska, the Pacific Northwest, and Brazil. He is also a noted lecturer on environmental leadership and the role of science in society.

Bradford S. Gentry is director of the Program on Private International Finance and the Environment at the Yale Center for Environmental Law and Policy. A practicing attorney, he spent seven years in London working on Euro-pean environmental issues and is an expert on structuring international infra-structure projects.

Bruce Guile is managing director of Washington Advisory Group, a con-sulting firm involved with public and private research and development efforts. He was previously director of the Program Office of the National Academy of Engineering, where he examined technological opportunities for improving environmental quality.

James K. Hammitt is associate professor of health policy and management in the Center for Risk Analysis at the Harvard School of Public Health. He is also director of the Environmental Science and Risk Management Program, which offers formal training in the application of quantitative methods to envi-ronmental policy.

Charles W. Powers is president of the Institute for Responsible Manage-ment. He previously served as vice president for public policy at Cummins Engine Company and was founding CEO of five public-private institutions, including Clean Sites, Inc., a pioneer in the use of dispute resolution in haz-ardous waste site clean-up.

John T. Preston is president and CEO of Quantum Energy, a high-technol-ogy company that develops energy-saving technologies. For ten years he led the office in charge of technology licensing and new venture spin-off at MIT. He is currently on the boards of directors of Molten Metal Technology, Clean Har-bors, and Energy BioSystems.

Carol M. Rose is Gordon Bradford Tweedy Professor of Law and Organiza-tion at the Yale Law School. Her writing and teaching fields include natural resources law, energy policy, land use regulation, public land management, water law, and the history and theory of property.

C. Ford Runge is professor of applied economics at the University of Minnesota. During 1987–1988, he served as special assistant to the Deputy U.S. Trade Representative in Geneva, and from 1988 to 1990 he served as the first director of the Center for International Food and Agricultural Policy at the University of Minnesota.

Jason Rylander is a freelance writer and environmental journalist. He is the managing editor of *Land Letter,* a biweekly environmental policy newsletter for natural resource professionals covering land use issues in Congress and the federal agencies from a national perspective. *Land Letter* is a publication of the Conservation Fund.

Stephan Schmidheiny is a Swiss industrialist whose companies range from high technology to retailing to forestry. He founded and chaired the Business Council for Sustainable Development, and served as principal advisor on business and industry to the Secretary General of the United Nations Conference on Environment and Development in 1992.

Robert Stavins is associate professor of public policy at the John F. Kennedy School of Government at Harvard University. He has written widely on the implementation of incentive-based approaches to environmental protection. He directed Project 88, the effort led by Senators John Heinz and Tim Wirth to advance the use of market mechanisms in environmental policy.

Todd Strauss is assistant professor of public policy and management science at the Yale School of Management. His research focuses on energy policy, including response by electric utilities and coal suppliers to the Clean Air Act, policy aspects of conservation and renewable fuels, electricity production and pricing, and utility regulation.

John Turner is president and CEO of the Conservation Fund in Arlington, Virginia. He served in the Wyoming State Legislature for eighteen years and was appointed director of the U.S. Fish and Wildlife Service in 1989. He joined the Conservation Fund in 1993 and directs its land conservation efforts.

John Urquhart is vice chairman of the board of Enron Corporation and president of John A. Urquhart Associates. In addition, he serves on the boards of the Aquarion Company, TECO Energy, Inc., Tampa Electric Company, Hubbel Incorporated, and the Weir Group, PLC. He is a retired officer of General Electric.

Bradley Whitehead is director of the Environmental Division of McKinsey and Company, the international management consulting firm. At McKinsey, he has helped corporations in a wide variety of industries develop and implement environmental strategies. He was also a contributor to Project 88, working with Robert Stavins.

Abbreviations

ASTM	American Society for Testing and Materials
BCA	Benefit-cost analysis
BLM	Bureau of Land Management
CERCLA	Comprehensive Environmental Response, Compensation, and Liability Act
CEA	Cost-effectiveness analysis
CFCS	Chlorofluorocarbons
CO	Carbon monoxide
CO_2	Carbon dioxide
CRA	Comparative-risk analysis
CRP	Conservation Reserve Program
DFE	Design-for-Environment
EMAS	Eco-management and audit scheme of the European Commission
EPA	U.S. Environmental Protection Agency
FAO	U.N. Food and Agriculture Organization
FASB	Financial Accounting Standards Board
FDI	Foreign direct investment
FERC	Federal Energy Regulatory Commission
FTC	Federal Trade Commission
GAO	General Accounting Office
GATT	General Agreement on Tariffs and Trade
GDP	Gross domestic product
GIS	Geographical information systems
GNP	Gross national product
HOT	High-occupancy or toll
HOV	High-occupancy vehicle
ISO	International Organization for Standardization
LCA	Life-cycle analysis
MAI	Multilateral Agreement on Investment
NAE	National Academy of Engineering
NAFTA	North American Free Trade Agreement
NAPA	National Academy of Public Administration
NEPA	National Environmental Policy Act
NEPI	National Environmental Policy Institute

NGO	Nongovernmental organization
NIHE	National Institutes of Health and Environment
NO_x	Nitrogen oxides
NRC	National Research Council
OECD	Organization for Economic Cooperation and Development
OPIC	Overseas private investment corporation
OSHA	Occupational Safety and Health Administration
OTA	U.S. Office of Technology Assessment
PCSD	President's Council on Sustainable Development
POTWS	Publicly-owned treatment works
RCRA	Resource Conservation and Recovery Act
RECLAIM	Regional Clean Air Incentives Market Program
RTA	Risk-tradeoff analysis
SCAQMD	South Coast Air Quality Management District
SO_2	Sulfur dioxide
SO_x	Sulfur oxide
TQM	Total quality management
TRI	Toxic release inventory
UNCTAD	U.N. Conference on Trade and Development
UNEP	U.N. Environment Programme
USDA	U.S. Department of Agriculture
VOCS	Volatile organic compounds
WTO	World Trade Organization

For Further Reading

Thinking Ecologically: An Introduction

Dowie, Mark. *Losing Ground: American Environmentalism at the Close of the Twentieth Century.* Cambridge: MIT Press, 1995.

Easterbrook, Gregg. *A Moment on the Earth.* New York: Viking, 1995.

Howard, Philip K. *The Death of Common Sense: How Law Is Suffocating America.* New York: Random House, 1994.

National Academy of Public Administration (NAPA). *Setting Priorities, Getting Results: A New Direction for the Environmental Protection Agency.* Washington, D.C.: NAPA, 1995.

National Commission on the Environment. *Choosing a Sustainable Future.* Washington, D.C., and Covelo, Calif.: Island Press, 1993.

President's Council on Sustainable Development (PCSD). *Sustainable America: A New Consensus for the Future.* Washington, D.C.: Government Printing Office, February 1996.

Schmidheiny, Stephan, with the Business Council for Sustainable Development. *Changing Course: A Global Business Perspective on Development and the Environment.* Cambridge: MIT Press, 1992.

World Commission on Environment and Development. *Our Common Future.* Oxford and New York: Oxford University Press, 1987.

Chapter 1. Industrial Ecology: Overcoming Policy Fragmentation

Allenby, B. R., and D. J. Richards, eds. *The Greening of Industrial Ecosystems.* Washington, D.C.: National Academy Press, 1994.

Ayres, Robert U., and Leslie W. Ayres. *Industrial Ecology: Towards Closing the Materials Cycle.* London: Edward Elgar, 1996.

Graedel, T. E., and B. R. Allenby. *Industrial Ecology.* Englewood Cliffs, N.J.: Prentice-Hall, 1995.

Journal of Industrial Ecology. Cambridge: MIT Press (initiated Spring 1997).

Lowe, Ernest, and John L. Warren. *The Source of Value: An Executive Briefing Sourcebook on Industrial Ecology.* Richland, Wash.: Pacific Northwest National Laboratory, 1996.

Socolow, Robert, et al., eds. *Industrial Ecology and Global Change.* Cambridge: Cambridge University Press, 1994.

Chapter 2. Ecosystem Management and Economic Development

Berry, J. K., A. A. Dyer, and R. Wu. *A Framework for Ecosystem Management in the Colorado Front Range: Incorporating the Human Dimension.* Final Report to USDA Rocky Mountain Forest and Range Experiment Station, Fort Collins, Colo., 1995.

Clark, Tim W. "Learning as a Strategy for Improving Endangered Species Conservation." Parts 1 and 2. *Endangered Species Update* 13, nos. 1 and 2 (1995).

Gunderson, L. H., C. S. Holling, and S. S. Light, eds. *Barriers and Bridges to the Renewal of Ecosystems and Institutions.* New York: Columbia University Press, 1995.

Kohm, Kathryn A., and Jerry F. Franklin, eds. *Creating a Forestry for the Twenty-first Century: The Science of Ecosystem Management.* Washington, D.C.: Island Press, 1997.

Science and Endangered Species Preservation: Rethinking the Environmental Policy Process. Special Report of the New York Academy of Sciences, Science Policy Program, G. Jon Rousch, moderator. August 1995.

Van Dyne, George M., ed. *The Ecosystem Concept in Natural Resource Management.* New York: Academic Press, 1969.

Vogt, Kristiina A., et al. *Ecosystems: Balancing Science with Management.* New York: Springer, 1997.

Chapter 3. Property Rights and Responsibilities

Bosselman, Fred, David Callies, and John Banta. *The Taking Issue.* Washington, D.C.: Government Printing Office, 1973.

Byrne, J. Peter. "Ten Arguments for the Abolition of the Regulatory Takings Doctrine." *Ecology Law Journal* 22 (1995): 89.

Ellickson, Robert C. "Property in Land." *Yale Law Journal* 102 (1993): 1315 .

Epstein, Richard. *Takings.* Cambridge: Harvard University Press, 1985.

Farber, Daniel. "Public Choice and Just Compensation." *Constitutional Commentary* 9 (1992): 279.

Fischel, William A. *Regulatory Takings: Law, Economics, and Politics.* Cambridge: Harvard University Press, 1995.

Freyfogle, Eric T. "The Owning and Taking of Sensitive Lands." *UCLA Law Review* 43 (1995): 77.

Halper, Louise A. "Why the Nuisance Knot Can't Undo the Takings Muddle." *Indiana Law Review* 28 (1995): 329.

Hart, John F. "Colonial Land Use Law and Its Significance for Modern Takings Doctrine." *Harvard Law Review* 109 (1996): 1252 .

Huffman, James. "Markets, Regulation and Environmental Protection." *Montana Law Review* 55 (1994): 425.

Libecap, Gary. *Contracting for Property Rights.* Cambridge and New York: Cambridge University Press, 1989.

Michelman, Frank I. "Property, Utility, and Fairness: Comments on the Ethical Foundation of 'Just Compensation' Law." *Harvard Law Review* 80 (1967): 1165.

Plater, Zygmunt. "Environmental Law as a Mirror of the Future." *Boston College Environmental Affairs Law Review* 23 (1996): 733 .

Rieser, Alison. "Ecological Preservation as a Public Property Right." *Harvard Environmental Law Review* 15 (1991): 393 .

Rose, Carol M. "A Dozen Propositions on Private Property, Public Rights, and the New Takings Legislation." *Washington and Lee Law Review* 53 (1996): 265.

———. "Takings, Federalism, Norms." *Yale Law Journal* 105 (1996).

Stanford Law Review 45 (1993): 1369–1455. Symposium on *Lucas v. So. Carolina Coastal Council.* Articles by Richard Epstein, William Fisher III, Richard Lazarus, and Joseph Sax.

Steinberg, Theodore. *Nature Incorporated: Industrialization and the Waters of New England.* Cambridge and New York: Cambridge University Press, 1991.

Stewart, Richard. "Environmental Regulation and International Competitiveness." *Yale Law Journal* 102 (1993): 2039.

Thompson, Edward Jr. *"Takings and Givings": Writing on Property Rights, Government Influence and Natural Resources.* Washington, D.C.: American Farmland Trust, 1996.

Treanor, William M. "The Original Understanding of the Takings Clause and the Political Process." *Columbia Law Review* 195 (1995): 782.

Underkuffler-Freund, Laura. "Takings and the Nature of Property." *Canadian Journal of Law and Jurisprudence* 9 (1996): 161.

Chapter 4. Land Use: The Forgotten Agenda

Bank of America et al. *Beyond Sprawl: New Patterns of Growth to Fit the New California.* Joint Report of the California Resources Agency, the Green Belt Alliance, the Low-Income Housing Fund, and the Bank of America. February 1995.

Diamond, Henry L., and Patrick F. Noonan. *Land Use in America.* Washington, D.C.: Island Press, 1996.

Endicott, Eve. *Land Conservation through Public/Private Partnerships.* Washington, D.C.: Island Press, 1993.

Hylton, Thomas. *Save Our Land, Save Our Towns.* Harrisburg, Penn.: RB Books, 1995.

Kunstler, James Howard. *The Geography of Nowhere.* New York: Touchstone, 1994.

Langdon, Philip. *A Better Place to Live: Reshaping the American Suburb.* New York: HarperCollins, 1994.

Leopold, Aldo. *A Sand County Almanac.* New York: Oxford University Press, 1949.

Platt, Rutherford H. *Land Use and Society.* Washington, D.C.: Island Press, 1996.

Chapter 5. Sorting out a Service-Based Economy

American Society for Healthcare Environmental Services. *An Ounce of Prevention: Waste Reduction Strategies for Healthcare Facilities.* Chicago: American Hospital Association, 1993.

Guile, B., and J. B. Quinn, eds. *Managing Innovation: Cases from the Services Industries.* Washington, D.C.: National Academy Press, 1988.

Hopkins, Lynne, David T. Allen, and Mike Brown. "Quantifying and Reducing Environmental Impacts Resulting from the Transportation of a Manufactured Garment." *Pollution Prevention Review* 4 (1994): 491–500.

Murphy, Paul R., Richard F. Poist, and Charles D. Braunschwieg. "Management of Environmental Issues in Logisitics: Current Status and Future Potential." *Transportation Journal* 34, no. 1 (22 Sept. 1994): 48.

Stahel, Walter R. "The Utilization-Focused Service Economy: Resource Efficiency and Product-Life Extension." In Allenby and Richards, *Greening of Industrial Ecosystems,* 178–90.

Wagner, Kathy, ed. *Environmental Management in Healthcare Facilities.* Philadelphia: W. B. Saunders, forthcoming.

Chapter 6. Globalization, Trade, and Interdependence

Anderson, Kym, and Richard Blackhurst, eds. *The Greening of World Trade Issues.* Ann Arbor, Mich.: University of Michigan Press, 1992.

Chayes, Abram, and Antonia Handler Chayes. *The New Sovereignty.* Cambridge: Harvard University Press, 1995.

Cleveland, Harlan. *Birth of a New World.* San Francisco: Jossey-Bass, 1993.

Commission on Global Governance. *Our Global Neighborhood.* New York: Oxford University Press, 1995.

Esty, Daniel C. *Greening the GATT: Trade, Environment, and the Future.* Washington, D.C.: Institute for International Economics, 1994.

Sand, Peter. *Lessons Learned in Global Environmental Governance.* Washington, D.C.: World Resources Institute, 1990.

Stone, Christopher D., *The Gnat Is Older than Man: Global Environment and Human Agenda.* Princeton: Princeton University Press, 1993.

Tay, Simon S. C., and Daniel C. Esty, eds. *Asian Dragons and Green Trade.* Singapore: Times Academic Press, 1996.

von Moltke, Konrad. "Why UNEP Matters." In *Green Globe Yearbook*. Oxford: Oxford University Press, 1996.

Chapter 7. Market-Based Environmental Policies

Hahn, Robert, and Robert Stavins. "Incentive-Based Environmental Regulation: A New Era from an Old Idea?" *Ecology Law Quarterly* 18 (1991): 1–42.

Organization for Economic Cooperation and Development (OECD). *Managing the Environment: The Role of Economic Instruments*. Paris: OECD, 1994.

Repetto, Robert, et al. *Green Fees: How a Tax Shift Can Work for the Environment and the Economy*. Washington, D.C.: World Resources Institute, 1993.

Stavins, Robert, ed. *Project 88—Harnessing Market Forces to Protect Our Environment: Initiatives for the New President*. Public Policy Study sponsored by Sen. Timothy E. Wirth (Colo.) and Sen. John Heinz (Pa.). Washington, D.C.: Government Printing Office, December 1988.

U.S. Environmental Protection Agency. Office of Policy, Planning, and Evaluation. *Economic Incentives: Options for Environmental Protection*. Washington, D.C.: Government Printing Office, 1991.

U.S. Congress, Office of Technology Assessment. *Environmental Policy Tools: A User's Guide*. OTA-ENV-634. Washington, D.C.: Government Printing Office, 1995.

Walley, Noah, and Bradley Whitehead. "It's Not Easy Being Green." *Harvard Business Review* 72, no. 3 (May–June 1994): 46–52.

Chapter 8. Privately Financed Sustainable Development

Asian Development Bank. *Financing Environmentally Sound Development*. Manila: Asian Development Bank, 1994.

Cairncross, Frances. *Costing the Earth*. London: Business Books, 1991.

Elkington, John, and Tom Burke. *The Green Capitalists: How to Make Money—and Protect the Environment*. London: Gollancz, 1989.

Esty, Daniel C., and Bradford S. Gentry. *Foreign Investment, Globalization, and the Environment*. Proceedings of the OECD Expert Workshop on Globalization and the Environment, Vienna, 30–31 Jan. 1997.

Gentry, Bradford S. *Private Investment and the Environment*. Discussion Paper 11. New York: Office of Development Studies, United Nations Development Programme, 1997.

Schmidheiny, Stephan, and Federico Zorraquin, with the World Business Council for Sustainable Development. *Financing Change: The Financial Community, Eco-efficiency, and Sustainable Development*. Cambridge: MIT Press, 1996.

Chapter 9. Technology Innovation and Environmental Progress

Ashford, Nicholas A. "An Innovation-Based Strategy for the Environment." In *Worst Things First? The Debate over Risk-Based National Environmental Priorities,* ed. A. M. Finkel and D. Golding. Washington, D.C.: Resources for the Future, 1994.

Heaton, George, Robert Repetto, and Rodney Sobin. *Transforming Technology: An Agenda for Environmentally Sustainable Growth in the Twenty-first Century.* Washington, D.C.: World Resources Institute, April 1991.

Moore, Curtis, and Alan Miller. *Green Gold: Japan, Germany, the United States, and the Race for Environmental Technology.* Boston: Beacon Press, 1994.

National Science and Technology Council. *National Environmental Technology Strategy.* Washington, D.C.: Interagency Environmental Technologies Office, 1995.

Nelson, Richard R. *Understanding Technical Change as an Evolutionary Process.* Amsterdam: Elsevier Science, 1987.

U.S. Congress, Office of Technology Assessment. *Industry, Technology, and the Environment: Competitive Challenges and Business Opportunities.* OTA-ITE-586. Washington, D.C.: Government Printing Office, January 1994.

Utterback, James M. *Mastering the Dynamics of Innovation: How Companies Can Seize Opportunities in the Face of Technological Change.* Boston: Harvard Business School Press, 1994.

Chapter 10. Data, Risk and Science: Foundations for Analysis

Breyer, Stephen. *Breaking the Vicious Circle: Toward Effective Risk Regulation.* Cambridge: Harvard University Press, 1993.

EPA, Office of Policy Analysis. *Unfinished Business: A Comparative Assessment of Environmental Problems.* Washington, D.C.: Government Printing Office, 1987.

EPA, Science Advisory Board. *Reducing Risk: Setting Priorities and Strategies for Environmental Protection.* Washington, D.C.: Government Printing Office, 1990.

Finkel, A. M., and D. Golding, eds. *Worst Things First? The Debate over Risk-Based National Environmental Priorities.* Washington, D.C.: Resources for the Future, 1994.

Graham, J., and J. Wiener, eds. *Risk versus Risk: Trade-offs in Protecting Health and the Environment.* Cambridge: Harvard University Press, 1995.

Hornstein, D. "Reclaiming Environmental Law: A Normative Critique of Comparative Risk Analysis." *Columbia Law Review* 92 (1992).

Office of Management and Budget. "Reforming Regulation and Managing Risk-Reduction Sensibly." In *Budget of the United States Government, Fiscal Year 1992.* Washington, D.C.: Government Printing Office, 1991.

Wargo, John. *Our Children's Toxic Legacy*. New Haven: Yale University Press, 1996.

Chapter 11. Toward Ecological Law and Policy

Ackerman, B., and W. Hassler. *Clean Coal/Dirty Air: Or How the Clean Air Act Became a Multibillion-Dollar Bail-out for High-Sulfur Coal Producers and What Should Be Done About It*. New Haven: Yale University Press, 1991.

Adler, Robert W., Jessica C. Landman, and Diane M. Cameron. *The Clean Water Act Twenty Years Later*. Washington, D.C. and Covelo, Calif.: Island Press, 1993.

Campbell-Moh, Celia, Barry Breen, J. William Futrell, and the Environmental Law Institute, eds. *Sustainable Environmental Law*. St. Paul, Minn.: West, 1993.

Carnegie Commission on Science, Technology, and Government. "Risk and the Environment: Improving Regulatory Decisionmaking." Washington, D.C., June 1993.

Cohen, S. "EPA: A Qualified Success." In *Controversies in Environmental Policy*, ed. Sheldon Kamieniecki, Robert O'Brien, and Michael Clarke. Albany: SUNY Press, 1986.

Elliott, E. Donald. "Environmental Law at a Crossroad." *North Kentucky Law Review* 20 (1992): 1–21 (Siebenthal Lecture).

Landy, M., M. Roberts, and S. Thomas. *The Environmental Protection Agency: Asking the Wrong Questions*. New York: Oxford University Press, 1990.

Melnick, R. S. *Regulation and the Courts: The Case of the Clean Air Act*. Washington, D.C.: NAPA, 1983.

Reilly, W. K. "The Future of Environmental Law." *Yale Journal of Regulation* 6 (1989): 351.

Shabecoff, P. *A Fierce Green Fire: The American Environmental Movement*. New York: Hill and Wang, 1993.

Chapter 12. Coexisting with the Car

Cronon, William. *Nature's Metropolis: Chicago and the Great West*. New York: W. W. Norton, 1991.

Downs, Anthony. *Stuck in Traffic: Coping with Peak-Hour Traffic Congestion*. Washington, D.C.: Brookings Institution, 1992.

Gordon, Deborah. *Steering a New Course: Transportation, Energy, and the Environment*. Washington, D.C.: Island Press, 1991.

Lovins, Amory, John W. Barnett, and L. Hunter Lovins. *Supercars: The Coming Light Vehicle Revolution*. Snowmass, Colo.: Rocky Mountain Institute, March 1993.

MacKenzie, James, Roger Dower, and Donald T. Chen. *The Going Rate: What it Really Costs to Drive*. Washington, D.C.: World Resources Institute, 1992.

Orski, C. Kenneth, ed. *Innovation Briefs.* New York, Urban Mobility Corp.

Samuel, Peter. *Highway Aggravation: The Case for Privatizing the Highways.* CATO Institute Policy Analysis 231. Washington, D.C.: CATO Institute, 27 June 1995.

U.S. Congress, Office of Technology Assessment (OTA). *Advanced Automotive Technology: Visions for a Super-Efficient Family Car.* Washington, D.C.: Government Printing Office, 1995.

Chapter 13. Fostering Environmental Policy from Farm to Market

Altieri, M. A., et al. *Agroecology: The Science of Sustainable Agriculture.* Boulder, Colo.: Westview Press, 1995.

Cochrane, Willard W., and C. Ford Runge. *Reforming Farm Policy: Toward a National Agenda.* Ames: Iowa State University Press, 1992.

Faeth, Paul. *Make It or Break It: Sustainability and the U.S. Agricultural Sector.* Washington, D.C.: World Resources Institute, 1996.

National Research Council (NRC), Committee on Pesticides in the Diets of Infants and Children, Board on Agriculture and Board on Environmental Studies and Toxicology, and Commission on Life Sciences. *Pesticides in the Diets of Infants and Children.* Washington, D.C.: National Academy Press, 1993.

Runge, C. Ford. "Agriculture and Environmental Policy: New Business or Business as Usual?" Working Paper no. 1. Environmental Reform: The Next Generation Project. Yale Center for Environmental Law and Policy, New Haven, September 1996.

U.N. Food and Agriculture Organization (FAO). *Livestock and the Environment: Finding a Balance.* Draft Summary Report for Steering Committee Approval, 19 April 1996.

U.S. Congress, Office of Technology Assessment. *Targeting Environmental Priorities in Agriculture: Reforming Program Strategies.* Washington, D.C.: Government Printing Office, October 1995.

Chapter 14. Energy Prices and Environmental Costs

Chandler, William U., ed. *Carbon Emissions Control Strategies: Case Studies in International Cooperation.* Washington, D.C.: World Wildlife Fund and Conservation Foundation, 1990.

Ivola, Pietro S., and Robert W. Crandall. *The Extra Mile: Rethinking Energy Policy for Automotive Transportation.* Washington, D.C.: Brookings Institution, 1995.

Landsberg, Hans H. *Making National Energy Policy.* Washington, D.C.: Resources for the Future, 1993.

Mintzer, Irving M., ed. *Confronting Climate Change: Risks, Implications, and Responses.* Cambridge: Cambridge University Press, 1992.

Brennan, Timothy J., et al. *A Shock to the System: Restructuring America's Electricity Industry*. Washington, D.C.: Resources for the Future, 1996.

U.S. Congress, Office of Technology Assessment. *Saving Energy in U.S. Transportation*. OTA-ETI-589. Washington, D.C.: Government Printing Office, July 1994.

Verleger, Philip K., Jr. *Adjusting to Volatile Energy Prices*. Washington, D.C.: Institute for International Economics, 1993.

Index

Adirondack Mountains, 42
Administrative Procedures Act, 163
agriculture, 13, 200–216; areas of opportunity
    in, 202–03; environmental impacts and,
    203–08; history of policy and, 200–203;
    mass-flow analysis and, 27–28;
    recommended policy ideas and, 212–15
AHMSA. *See* Altos Hornos de Mexico S.A.
air quality. *See also* automobiles; Clean Air
    Act and, 11, 20, 57, 108, 109, 159–60;
    energy use and, 218, 223–29
Allenby, B., 25–26
Altos Hornos de Mexico S.A. (AHMSA), 122,
    126
Anderson, F., 36n33
Association of South East Asian Nations
    (ASEAN), 100
automobiles, 13, 189–99; cost incentives and,
    194–98, 236; design of, 192; driver
    behavior and, 190–91; environmental
    technologies and, 137, 139, 219; fuel
    technology and, 192, 199n2, 222–23;
    gasoline taxes and, 221–23; habitat
    destruction and, 231–32; information
    technology and, 193; mass transportation
    and, 193–94; toll roads and, 195–98;
    transportation policy and, 64, 192, 193–95
avoidance technologies, 139

barriers to change, 22, 23, 30; approaches to
    dealing with, 30–33, 111–15;
    implementation of market mechanisms
    and, 109–11, 116n11
Baxter Healthcare S.A., 125
BCA. *See* benefit-cost analysis
benefit-cost analysis (BCA), 69, 152, 155–56,
    160
big dirties, 172, 219
biodiversity, 4, 70, 99, 118, 121, 206–08

"bottle bills," 108
Breyer, S., 162
Brewer, G., 5
Brookings Institution, 190
Brownstein, M., 33n4
Brunner, R., 5, 36n34
Bryson, J., 2–3
"bubble" concept, 179–83

California, 108, 147, 192, 208, 224; Safe
    Drinking Water and Toxic Enforcement
    Act of *1986,* 164; toll roads projects in,
    197, 198
cap-and-trade system. *See* "bubble" concept
carbon dioxide, 115, 140–41, 158, 199, 218,
    222, 227
Cato Institute, 190
CEA. *See* cost-effectiveness analysis
CERCLA, 34n8
certification: international, 122; programs,
    95, 131; self- and third-party, 176, 177
Charnovitz, S., 11, 91–102
chemical industry, 111, 137, 143
Chertow, M. R., 1–16, 9, 19–36, 183,
    231–40
China, 12
Clark, T., 5, 36n34
Clean Air Act, 20, 159–60, 171, 179;
    Amendments of *1990,* 11, 57, 108, 109,
    174, 180, 225
Clean Water Act, 3, 20, 201
climate change, 28, 99, 114, 116n5, 153,
    158, 218
Coase, R., 116n7, 181
Coastal Barriers Resources Act, 68
Coastal Zone Management Act, 68
Cohon, J., 10, 76–90
collaborative approaches: ecosystem
    management and, 41–43, 44, 45–46;

Thinking Ecologically
The Next Generation of Environmental Policy
Edited by Marian R. Chertow and Daniel C. Esty

Twenty-five years ago, the Cuyahoga River in Ohio was so contaminated that it caught fire, air pollution in some cities was thick enough to taste, and environmental laws focused on the obvious enemy: large American factories with belch smokestacks and pipes gushing wastes. Federal legislation has succeeded in providing cleaner air and water, but we now confront a different set of environmental problems—less visible and more subtle. This important book offers thought-provoking ideas on how America can respond to changing public health and ecological risks and create sound environmental policy for the future.

The innovative thinkers of the Next Generation Project of the Yale Center for Environmental Law and Policy—experts from business, government, nongovernmental organizations, and academia—propose reforms that balance environmental efforts with other public needs and issues. They call for new foundations for environmental law and policy, adoption of a more diverse set of policy tools and strategies (economic incentives, eco-labels), and new connections between critical sectors (agriculture, energy, transportation, service providers) and environmental policy. Further progress must involve not only officials from the U.S. Environmental Protection Agency and state environmental protection departments, say the authors, but also decisionmakers as diverse as mayors, farmers, energy company executives, and delivery route planners. To be effective, next-generation policymaking will view environmental challenges comprehensively, connect academic theory with practical policy, and bridge the gaps that have caused recent policy debates to break down in rancor. This book begins the process of accomplishing these challenging goals.

Marian R. Chertow is director of the Next Generation Project and a member of the faculty of the Yale School of Forestry and Environmental Studies. **Daniel C. Esty** is director of the Yale Center for Environmental Law and Policy and has a joint appointment at Yale's School of Law and School of Forestry and Environmental Studies.

A Yale Fastback

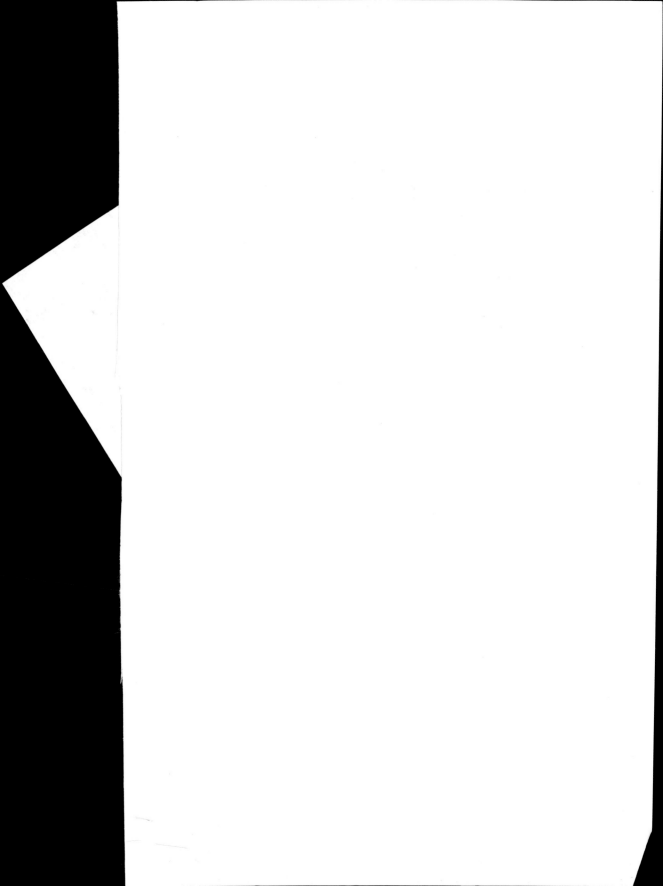

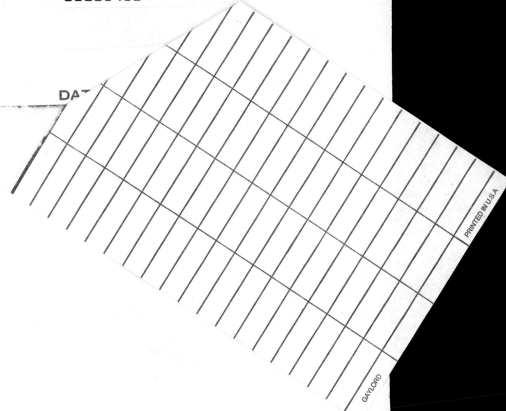

DA